Laboratory Manual
for Physical Geology

Laboratory Manual for Physical Geology

Sixth Edition

Norris W. Jones
Emeritus Professor, University of Wisconsin–Oshkosh

Charles E. Jones
University of Pittsburgh

Featuring the Photography of Doug Sherman and Dr. Parvinder Sethi

 Higher Education

Boston   Burr Ridge, IL   Dubuque, IA   New York   San Francisco   St. Louis
Bangkok   Bogotá   Caracas   Kuala Lumpur   Lisbon   London   Madrid   Mexico City
Milan   Montreal   New Delhi   Santiago   Seoul   Singapore   Sydney   Taipei   Toronto

Higher Education

LABORATORY MANUAL FOR PHYSICAL GEOLOGY, SIXTH EDITION

Published by McGraw-Hill, a business unit of The McGraw-Hill Companies, Inc., 1221 Avenue of the Americas, New York, NY 10020. Copyright © 2008 by The McGraw-Hill Companies, Inc. All rights reserved. No part of this publication may be reproduced or distributed in any form or by any means, or stored in a database or retrieval system, without the prior written consent of The McGraw-Hill Companies, Inc., including, but not limited to, in any network or other electronic storage or transmission, or broadcast for distance learning.

Some ancillaries, including electronic and print components, may not be available to customers outside the United States.

 This book is printed on recycled, acid-free paper containing 10% postconsumer waste.

2 3 4 5 6 7 8 9 0 QPD/QPD 0 9 8

ISBN 978–0–07–721894–2
MHID 0–07–721894–9

Publisher: *Thomas D. Timp/Margaret J. Kemp*
Senior Developmental Editor: *Margaret B. Horn*
Project Manager: *Joyce Watters*
Senior Production Supervisor: *Laura Fuller*
Lead Media Project Manager: *Judi David*
Media Producer: *Daniel M. Wallace*
Designer: *David W. Hash*
(USE) Cover Image: *Lava flows into the Sea at Hawaii during the eruption of Kilauea, ©Parvinder Sethi*
Senior Photo Research Coordinator: *Lori Hancock*
Compositor: *Carlisle Publishing Services*
Typeface: 10/12 *Times Roman*
Printer: *Quebecor World Dubuque, IA*

The credits section for this book begins on page 343 and is considered an extension of the copyright page.

Contents

List of Tables and Maps

Tables

Maps

Preface

To the Student

Geology and other sciences are major parts of our lives, and many of the daily decisions we make are affected in some way by science. Because of this, all educated people (scientists or not) should understand, in a general way, how science is done, and should be able to apply the basic ideas of the scientific method. Thus, a principal objective of this laboratory manual is to show you how geologists use the scientific method by having you do it yourselves. You will find that geologists, like other scientists, operate in a very logical and straightforward manner, which differs little from the way you normally do things. You will learn how to gather and use data (for example, making measurements to use in calculations), establish relations based on these data, classify information, develop and test hypotheses or models to explain observations and relations, and use this information to draw conclusions and make predictions. This may sound a bit intimidating, but don't panic; it's not difficult. You'll be surprised at what you'll be able to do.

Laboratory sessions will be most efficient if your **read the entire chapter before coming to class.** If allowed by the instructor, students may work in groups. In a group setting, it soon becomes apparent who has prepared and can contribute to the group. As a courtesy to others, and for your own benefit, *prepare for class!* Make sure you bring those things to class that are needed for the lab, as indicated in the **Materials Needed** section. Don't depend on your classmates.

You will find that geology is relevant to our daily lives. Look around you. Most of the manufactured items you see are made of earth materials—minerals, rocks, or petroleum. We live along rivers that flood, drink water pumped from underground, and dispose of our garbage in the ground. We and our politicians have to understand the consequences of our actions. So pay attention, ask questions, and learn to think critically about these geological issues. Make sure that science-related issues are based on good science; don't let charlatans determine our public policy.

To the Instructor

This lab manual is for a college-level, introductory course in physical geology. It is based on the following assumptions: (1) A typical student is a freshman or sophomore concurrently in a physical geology class. He or she has little background in science and math. The student may have had a unit on Earth science in seventh, eighth, or ninth grade, but most likely has had little or no exposure to geology. (2) The course has no prerequisites. (3) Although a few students taking the class plan to major in geology, and more will decide to major once they learn how interesting geology is, most students are taking the class to satisfy a general requirement and will not major in geology or other sciences.

Each chapter contains more exercises than your students may have time to do, so you will have to be selective as to which problems you assign. The sequence of topics is a traditional one, but the instructor is free to choose which topics to cover as well as the order in which they are covered. The text in each chapter is intentionally brief. It is meant to serve as background for the problems rather than a "complete" treatment of the subject. An Instructor's Manual is available as PDF and Word files on-line (*www.mhhe.com/jones6e*) and is accessible with a password obtained from the publisher. We strongly recommend you take a look at it.

The first edition of this manual, by NWJ, was the outgrowth of twenty-five years of teaching physical geology and using many different lab exercises. CEJ became a co-author and added his expertise to the fourth and subsequent editions. In this, the sixth edition, we offer two chapters that have been revised extensively. Chapter 6, Topographic Maps and Digital Elevation Models, now introduces digital elevation models and offers new problems that we hope are of more interest to the students. Both the text and the problems of the new Chapter 7 emphasize volcanoes, as indicated by the new title: Remote Sensing of Volcanoes and Volcanic Hazards.

With each new edition, we strive to increase clarity in both the text and problems. We also try to meet changing instructional needs by keeping both text and problems as current as is necessary. We are aided in this task by the very helpful comments and suggestions from our reviewers. If you find any errors in the manual or have suggestions for future improvements, **please visit the lab manual web page (*www.mhhe.com/jones6e*).** There you will find the most up-to-date web addresses for the WWW problems, a page listing any *errata* discovered following publication, and our e-mail addresses. Our goal is to create a lab manual that meets the needs of as many instructors as possible while still maintaining its relatively slender size.

Most of the photographs in this manual were taken by geology professors Doug Sherman and Dr. Parvinder Sethi, both of whom are also professional photographers. Their expertise in both areas made our job of selecting illustrations much easier.

To Everyone: Organization and Use

This manual consists of 17 chapters that cover the usual topics in physical geology. The first five chapters deal with *Earth Materials*—minerals and rocks. Because maps, aerial photographs, and other types of remotely sensed images are used so extensively in geology, two chapters focus on *Maps and Images,* especially as related to volcanoes. Five chapters then show how *Landscapes and Surface Processes* are studied using maps and images. One laboratory is devoted to Geologic Time and Sequences, and in the final chapters, Internal Processes and their effects on the surface are studied. Each chapter contains a list of materials needed to do the lab (**Materials Needed**); sufficient descriptive text so that all lab problems can be completed without reference to another textbook; a list of **Objectives,** which summarizes the things that a student will be expected to do or know after completing the lab; and laboratory **Problems** or exercises that are required

for the lab. The **Objectives** are located immediately before the **Problems**, rather than at the beginning of the chapter, because we want students to relate them to the problems more than the text.

The problems are designed to illustrate basic principles of geology and methods of solving geological problems. A paragraph preceding the objectives in each chapter briefly indicates how the problems relate to the scientific method. As often as possible, problems were chosen to demonstrate practical aspects. Some problems require making measurements and doing simple calculations. Most questions require specific, short answers. A few questions, however, are designed so that several answers may be possible. Students must formulate several hypotheses, then determine what additional information would be needed before a single hypothesis could be chosen. Problems that require more time to complete or are more difficult are preceded by the heading *In Greater Depth.* It is up to the individual instructor to determine how many problems, or what parts of individual problems, should be assigned in his or her class.

All chapters except Chapter 5 (Metamorphic Rocks) have a **Web Problem,** which can be solved from information available on the World Wide Web. We have provided specific web addresses for all but one problem, so they don't take an unreasonable amount of time. The student can access these websites by entering the address given in the individual Web Problem or by using the links provided by McGraw-Hill at *www.mhhe. com/jones6e*, linking to *Student Center*, and selecting the appropriate chapter. We encourage you to gain access through *www.mhhe.com/jones6e*, because of the changeable nature of websites. If changes occur that affect an individual problem, they will be noted and available at *www.mhhe.com/jones6e*.

The inside of the front cover of the manual contains a plate tectonic map of the world. The inside of the back cover lists symbols of selected chemical elements, units of measurement and factors for converting between metric and English units, and a brief explanation of the use of powers of ten to abbreviate large and small numbers. Key terms are in **boldface** in the text. Italics are used to call attention to a word or phrase. Answer pages can be torn out and handed in for grading if necessary. Colored, tear-out, structure models at the end of the manual are needed for the problems in Chapter 14.

Acknowledgements

NWJ would like to thank his colleagues at the University of Wisconsin–Oshkosh, Bill Fetter, Gene LaBerge, Tom Laudon, Jim McKee, Brian McKnight, and Bill Mode, for their comments, suggestions, and moral support over the years. CEJ especially thanks wife Dana, child prodigy Hannah, and adorable clones Maya and Abby for putting up with many hours of neglect during this revision. We also thank the following reviewers of the lab manual: Paul E. Bellah—Collin County Community College, Nathalie N. Brandes—Montgomery College, Jad DíAllura—Southern Oregon University, Christie Demosthenous—University of Wisconsin–Oshkosh, Pamela J. W. Gore—Georgia Perimeter College, Bernard Housen—Western Washington University, David R. Schwimmer—Columbus State University, Yuch-Ning Shieh—Purdue University, Mark J. Sutherland—College of DuPage, F. Colleen Wendeborn—Louisiana State University, and McGraw-Hill Developmental Editor Margaret Horn.

Prologue: Methods of Science

Science is a systematized body of knowledge derived mostly from observation and logical inference. The ultimate goal of scientists is to discover the basic principles that determine how things behave. Scientists have the following in mind as they work: (1) natural processes are orderly, consistent, and predictable; (2) there is a cause (or set of causes) for every effect; and (3) the best explanations are usually the simplest (*Doctrine of Simplicity or Parsimony*). Geologists and other scientists who study the past assume that physical and chemical laws, such as the law of gravity, are constant and have not varied over time (*Doctrine of Actualism*).

Scientific advances come in different ways. Some are based on theoretical ideas about things not yet observed (for example, Einstein's theory of relativity). Others result from unplanned, fortuitous events (such as the eruption of Mount St. Helens volcano). But most advances result from a painstaking approach known as the *scientific method*.

The **scientific method** generally involves: (1) formulation of a problem or question; (2) observation and measurement to determine characteristics or properties, to measure variation, classify, and establish relations—in other words, to get a basic understanding of the situation; (3) formulation of one or more explanations, models, or **hypotheses** to explain the observations; (4) experimentation and testing to reject or modify hypotheses; and (5) acceptance of a best hypothesis that withstands all tests and the scrutiny of peers as a **theory**.

The scientific method is *not* a rigid series of steps that must always be followed to do science. For example, most scientists already have a tentative hypothesis in mind when making observations. In fact, as philosopher Karl Popper said, successful scientists often are those with the best "imaginative preconception" of the correct hypothesis.

However, to avoid being blinded by such preconception, geologist T.C. Chamberlin recommended maintaining *multiple working hypotheses* until such time as they could be reduced to only one. With this approach, the scientific process involves rejecting those hypotheses that do not withstand testing and experimentation. Commonly, the simplest hypothesis that explains the most observations is the one that survives to become a theory. However, even though a theory is supported by a great deal of evidence and seems to explain all observations, it still is subject to testing, and eventually it may be rejected or modified as a result of new observations. Theories that withstand all testing and are invariable become **scientific laws;** an example is Newton's Law of Gravitation (the law of gravity).

The scientific method may seem rather involved, but it is an approach you probably use in many situations without realizing it. Take a very simple example. Let's say you go to geology lab one day and find nobody in the classroom. That observation should raise your curiosity. Why is no one there? Several possible explanations, or *hypotheses,* come to mind: (1) the professor is ill and the class was canceled; (2) no class was scheduled for that day; (3) you are there at the wrong time or on the wrong day; (4) the class has gone on a field trip that you forgot; (5) the class is hiding in the room, ready to leap out and yell "Happy birthday"; or (6) everyone has been lifted magically from the room. As a good scientist and rational thinker, you rank them in order of which is the most likely, and quickly eliminate those that are impossible or too unlikely. The last one (6) is not testable and is out of the realm of science, because it calls on a supernatural phenomenon for an explanation. A quick check of the time and date allows you to eliminate number 3. It's not your birthday, so you forget about number 5. Three reasonable hypotheses remain to be tested.

One way to test a hypothesis is to assume it is true, make a prediction based on that assumption, then see if the prediction comes true. If the professor is ill (hypothesis number 1), you might predict that a note to that effect would be posted on the door or the blackboard, but there is none. You might also expect to see other curious students about, but there are none. These negative results strongly suggest that hypothesis 1 is incorrect, although they do not necessarily prove it wrong. After all, it is possible that the professor was not able to have a notice put in the classroom, and it is possible that all the other students decided not to come to class that day. Those are pretty complex and improbable explanations, however. Hypotheses 2 and 4 are simpler and more reasonable. An easy way to test these is to look at the syllabus for the class.

Let's assume that no class is scheduled, according to the syllabus. Then you can establish a theory. The theory is that the classroom is empty because there is no class that day. Note that this is a *scientific theory* (although theories in science usually require much more time to develop), one that can be tested in many ways, with the same result. For example, you could check with the Geology Department office or look at another student's syllabus. The public tends to use the word theory in quite a different sense, almost disdainful, as something that has little basis in fact. As you can see from the example, a *scientific theory* is firmly established and as close to truth as scientifically possible. However, even a scientific theory is always subject to more testing and possible revision in light of new evidence.

That is how much of science works. There is nothing mysterious about it. It is simply a logical, open-minded approach to a question. Conclusions are not made without considerable thought and testing. It is a rational approach to all kinds of problems or questions, not just scientific ones, and undoubtedly is the approach you have used many times yourself. As you do the exercises in your geology laboratory, you will be using the scientific method as applied by geologists.

Part I
Minerals

Cubic crystals of pyrite at three times actual size.

CHAPTER 1

Properties of Minerals

Materials Needed

- Pencil
- Hand lens
- Calculator
- Set of mineral samples

- Equipment for determination of mineral properties: streak plate, copper penny, glass plate or steel knife, magnet, dilute hydrochloric acid, contact goniometer (optional)

Introduction

To understand most geologic processes, we must understand rocks. And to understand rocks we must first understand minerals, which are the components of rocks. Minerals differ from each other in chemical composition and atomic arrangement, and these factors produce distinctive physical properties that enable minerals to be identified. In this chapter you will learn how to determine or identify the most useful properties. In Chapter 2, these properties are used to identify minerals.

What Is a Mineral?

A **mineral** is a naturally occurring compound or chemical element made of atoms arranged in an orderly, repetitive pattern. Its chemical composition is expressed with a chemical formula (**symbols for important elements are given inside the back cover**). Both chemical composition *and* atomic arrangement characterize a mineral and determine its physical properties. Most minerals form by inorganic processes, but some, identical in all respects to inorganically formed minerals, are produced by organic processes (for example, the calcium carbonate in clam shells). A few naturally occurring substances called **mineraloids** have characteristic chemical compositions, but are amorphous; that is, atoms are *not* arranged in regular patterns. Opal is an example.

The precise chemical composition and internal atomic architecture that define each mineral also directly determine its outward appearance and physical properties. Thus, in most cases, general appearance and a few easily determined physical properties are sufficient to identify the mineral.

Physical Properties

Color, luster, streak, hardness, cleavage, fracture, and crystal form are the most useful physical properties for identifying most minerals. Other properties—such as reaction with acid, magnetism, specific gravity or density, tenacity, taste, odor, feel, and presence of striations—are helpful in identifying certain minerals.

Color

Color is the most readily apparent property of a mineral, but BE CAREFUL. Slight impurities or defects within the crystal structure determine the color of many minerals. For example, quartz can be colorless, white, pink, purple, green, gray, or black (Fig. 1.1). Color generally is diagnostic for minerals with a metallic luster (defined in next section) but may vary quite a bit in minerals with a nonmetallic luster. **Check the other properties before making an identification.**

Luster

Luster describes the appearance of a mineral when light is reflected from its surface.

Is it shiny or dull? Does it look like a metal or like glass? Most minerals have either a **metallic** or **nonmetallic** luster. As you will see, the first thing you must determine before a mineral can be identified using the tables in Chapter 2 is whether its luster is metallic or nonmetallic.

Minerals with a metallic luster look like a metal, such as steel or copper (Fig. 1.2A, B; Part 1 opener). They are opaque, even when looking at a thin edge. Many metallic minerals become dull looking when they are

× 0.2

FIGURE 1.1
Color is NOT a reliable criterion for the identification of many minerals, as illustrated here by three varieties of the mineral quartz: amethyst (purple), rock crystal (clear), and smoky (black). The × 0.2 below the picture is a shorthand notation to indicate the actual size of the mineral samples. It means that the samples are 0.2 times the size in the picture; that is, the picture shows them one-fifth (0.2) their actual size. This notation will be used throughout this lab manual.

A. Metallic (galena) × 0.3

B. Metallic (pyrite) × 0.5

C. Nonmetallic, vitreous (muscovite) × 0.5

D. Nonmetallic, vitreous (biotite) × 0.5

E. Nonmetallic, dull (kaolinite) × 0.5

F. Nonmetallic, pearly (talc) × 0.5

G. Nonmetallic, waxy (jasper) × 0.5

H. Nonmetallic, resinous (sphalerite) × 0.5

exposed to air for a long time (like silver, they tarnish). To determine whether or not a mineral has a metallic luster, therefore, you must look at a recently broken part of the mineral.

Minerals with nonmetallic luster can be any color. At first, you may have difficulty determining whether some black minerals have metallic or nonmetallic luster. However, thin pieces or edges of minerals with nonmetallic luster generally are translucent or transparent to light, and even thick pieces give you the sense that the reflected light has entered the mineral a bit before being reflected back. There are several types of nonmetallic lusters. *Vitreous luster* is like that of glass (Fig. 1.2C, D). Remember that glass can be almost any color, including black, so don't be fooled by the color. A *dull luster* has an earthy appear-

ance caused by weak or diffuse reflection of light (Fig. 1.2E). Other nonmetallic lusters include: *pearly luster,* like a pearl or the inside of a fresh clam shell (Fig. 1.2F); *greasy luster,* as though covered by a coat of oil; *waxy luster,* like paraffin (Fig. 1.2G); and *resinous luster,* like resin or tree sap (Fig. 1.2H).

Hardness

Hardness is the resistance of a smooth surface to abrasion or scratching. A harder mineral scratches a softer mineral, but a softer one does not scratch a harder one. To determine the hardness of a mineral, something with a known hardness is used to scratch, or be scratched by, the unknown. The minerals in the Mohs Hardness Scale (Table 1.1) are used as standards for com-

parison. The Mohs hardnesses are also given for some common items, which you can use to determine the hardness of an unknown mineral if you don't happen to have a pocketful of Mohs minerals.

To determine hardness, run a sharp edge or a point of a mineral with known hardness across a smooth face of the mineral to be tested (Fig. 1.3). Do not scratch back and forth like an eraser, but press hard and slowly scratch a line, like you are trying to etch a groove in glass. Make sure that the contact points of both minerals are the minerals you intend to test, and not impurities. Also, make sure that the mineral has actually been scratched. Sometimes powder of the softer mineral is left on the harder mineral and gives the appearance of a scratch on the harder one. Brush the tested surface with your finger to see if a groove or

Table 1.1
Scale of Hardness

Mohs Hardness Scale	Mohs Hardness of Common Items
10..Diamond	These usefully constrain hardnesses
9..Corundum	of many common minerals.
8..Topaz or Beryl	
7..Quartz	Streak plate ....6.5 to 7
6..Orthoclase feldspar	Steel knife or steel nail ......5 to 6
5..Apatite	Glass ....5 to 5.5
4..Fluorite	Copper coin 3.5
3..Calcite	Fingernail ....2 to 2.5
2..Gypsum	
1..Talc	

scratch remains. You may need to use a hand lens or magnifying glass to see whether a scratch was made. To double-check your results, try scratching the substance of known hardness with the mineral of unknown hardness. If two minerals have the same hardness, they may be able to scratch each other.

A piece of window glass is commonly used in geology labs as a standard for determining hardness. There are several reasons for this: (1) it's easy to see a scratch on glass; (2) the hardness of glass (5 to 5½) is midway on the Mohs scale; and (3) glass is inexpensive and easily replaced. In fact, you will discover that the hardness of a mineral when compared to glass is one of the principal bases in identifying a mineral using the determinative tables in Chapter 2.

Put the piece of glass on a stable, flat surface such as a tabletop. Then rub the mineral on the glass, and check to see if the glass was scratched. DO NOT TRY TO SCRATCH THE MINERAL WITH THE GLASS, because glass chips easily.

More sophisticated methods than the scratch test have been developed to determine hardness; Figure 1.4 illustrates how some of the minerals of the Mohs scale compare with their hardnesses, as determined by another method. As you can see, Friedrich Mohs, who devised the scale in 1822, did a pretty good job of selecting minerals so that the intervals between adjacent minerals on his scale were consistent.

Streak

Streak is the color of the mineral when finely powdered; it may or may not be the same color as the mineral. Streak is more helpful for identifying minerals with metallic lusters, because those with nonmetallic lusters generally have a colorless or light-colored streak that is not very diagnostic. Streak is obtained by scratching the mineral on an unpolished piece of white porcelain called a streak plate (Fig. 1.5). Because the streak plate is harder than most minerals, rubbing the mineral across the plate produces a powder of that mineral. When the excess powder is blown away, what remains is the color of the streak. Because the streak of a mineral is usually the same, no matter what the color of the mineral, streak is commonly more reliable than color for identification.

FIGURE 1.3
The harder white mineral (calcite) scratches the softer one (gypsum).

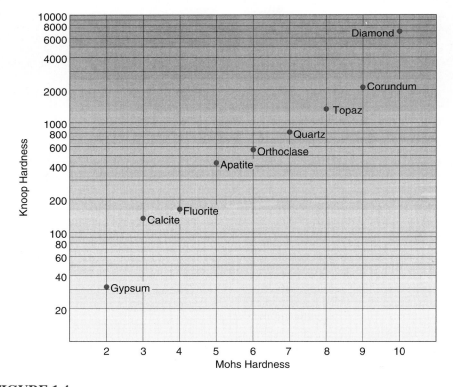

FIGURE 1.4
Comparison between Mohs-scale hardness and "actual" hardness (Knoop hardness). Note that "actual" hardness is shown with a logarithmic scale. Talc (Mohs hardness of 1) was not included in the study that produced these data.

FIGURE 1.5

The streak of this dark gray mineral (hematite), obtained by rubbing it on the white streak plate, is reddish brown.

Cleavage and Fracture

The way in which a mineral breaks is determined by the arrangement of its atoms and the strength of the chemical bonds holding them together. Breakage occurs where chemical bonds are relatively weak and where there are no atoms. Because these properties are unique to the mineral, careful observation of broken surfaces may aid in mineral identification. A mineral that exhibits **cleavage** consistently breaks, or *cleaves,* along parallel, flat surfaces called **cleavage planes.** A mineral **fractures** if it breaks along random, irregular surfaces. Some minerals break only by fracturing, while others both cleave and fracture.

The mineral halite (sodium chloride [NaCl], or salt) illustrates how atomic arrangement determines the way a mineral breaks. Figure 1.6 shows the arrangement of sodium and chlorine atoms in halite. Notice that there are planes with atoms and planes without atoms. When halite breaks, it breaks parallel to the planes with atoms but along the planes without atoms. Because there are three *directions* in which atom density is equal, halite has three **directions of cleavage,** each at 90° to each other. The *number of cleavage directions* and the angles between them are important in mineral identification because they reflect the underlying atomic architecture that helps define each mineral.

Cleavage planes, as flat surfaces, are easily spotted by turning a sample in your hand until you see a single flash of reflected light from across the mineral surface. Individual cleavage surfaces may extend across the whole mineral specimen (Fig. 1.7) or, more commonly, they may be offset from each other by small amounts, as illustrated in Figure 1.8. Even though they are offset, they work as tiny mirrors that create the single flash seen in Figure 1.9. Figures 1.10 and 1.11A also show offset cleavage surfaces.

Cleavage quality is described as *perfect, good,* and *poor.* Minerals with a *perfect* or excellent cleavage break easily along flat surfaces and are easy to spot. Minerals with *good* cleavages do not have such well-defined cleavage planes and reflect less light. *Poor* cleavages are the toughest to recognize, but can be spotted by small flashes of light in certain positions. All cleavages illustrated here are perfect or good.

Minerals have characteristic numbers of cleavages. This number is determined by counting the number of cleavage surfaces that are *not* parallel to each other. For example, the mineral in Figure 1.7 has two cleavage surfaces that are visible plus the one lying on the table. However, each of these cleavage surfaces is parallel to the other, so this mineral is said to have only **one cleavage direction.** Minerals with one cleavage are often said to have a *basal cleavage.*

Two cleavage directions are illustrated in Figure 1.10. In Figure 1.10A, the cleavage surfaces or planes are horizontal and vertical: they intersect each other at 90°. In Figure 1.10B, the cleavage surfaces are horizontal and steeply inclined: they do not intersect at 90°, but at 56° and 124°. The fronts and backs of each specimen are irregular fracture surfaces. Counting only *nonparallel cleavage surfaces,* each sample shows two cleavage directions, one horizontal and one vertical (Fig. 1.10A) or inclined (Fig. 1.10B). Intersecting cleavages may define an elongate geometric object

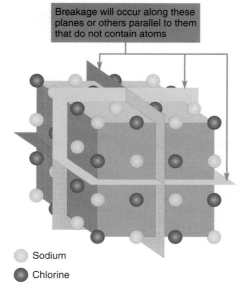

Breakage will occur along these planes or others parallel to them that do not contain atoms

○ Sodium
● Chlorine

×1

FIGURE 1.6

The illustration shows that atoms of sodium and chlorine in the mineral halite are parallel to three planes that intersect at 90°. Halite breaks, or *cleaves,* most easily between the three planes of atoms, so it has three *directions* of cleavage that intersect at 90°. The photograph of halite illustrates the three directions of cleavage.

One direction of cleavage

×1

FIGURE 1.7
Biotite mica has one direction of cleavage (basal cleavage).

×0.5

FIGURE 1.8
When incoming light rays (yellow lines) strike parallel cleavage surfaces (seen here in cross section), the rays are all reflected in the same direction (pink lines), even though the surfaces are at different elevations. When viewed from that direction, the mineral shines. Rays striking irregular fracture surfaces reflect in many different directions (green lines), causing fracture surfaces to appear duller.

FIGURE 1.9
The parallel cleavage planes in hornblende shine because the light is reflected in the same direction.

called a *prism;* such minerals are said to have a *prismatic cleavage.* When there are two cleavages, you should note the angle between them. Most commonly, cleavage angles are at or close to 90° (Fig. 1.10A) or 60° and 120° (Fig. 1.10B).

With **three cleavage directions,** a mineral can be broken in the shape of a cube if the three cleavages intersect at 90° (called a *cubic cleavage;* Fig. 1.11A) or a rhombohedron if the angles are not 90° (called a *rhombohedral cleavage;* Fig. 1.11B).

Minerals with **four** or **six cleavage directions** are not common. Four cleavage

planes can intersect to form an eight-sided figure known as an octahedron (Fig. 1.12). Fluorite is the most common mineral with an *octahedral cleavage.* Six cleavage directions intersect to form a *dodecahedron,* a twelve-sided form with diamond-shaped faces. A common mineral with *dodecahedal cleavage* is sphalerite (Fig. 1.13).

When counting cleavage directions it is essential that you count surfaces on just *one* mineral crystal. The photographs shown here use single, large, broken crystals to illustrate cleavage. In nature you often find that a single hand-sized sample

contains a large number of crystals grown together (see following discussion under "Crystal Form"). If you count up cleavage surfaces from more than one crystal, a wrong number is likely.

Finally, **fracture surfaces** can cut a mineral grain in any direction. Fractures are generally rough or irregular, rather than flat, and thus appear duller than cleavage surfaces (Fig. 1.14A). Some minerals fracture in a way that helps to identify them. For example, quartz has no cleavage but, like glass, it breaks along numerous small, smooth, curved surfaces called *conchoidal*

×1

Cleavage plane

Fracture surface

90°

Cleavage plane

A. Cleavages intersect at 90°

× 0.5

Cleavage plane

Fracture surface

56°   124°

Cleavage plane

B. Cleavages intersect at 56° and 124°

FIGURE 1.10
Two directions of cleavage: A. Cleavages in potassium feldspar intersect at 90°; B. Prismatic cleavages in hornblende intersect at 56° and 124°.

×1

Cleavage plane

Cleavage planes

A. All three cleavages intersect at 90°

× 0.8

Cleavage plane

Cleavage planes

~75°   ~105°

B. All three cleavages intersect at about 75°/105°

FIGURE 1.11
Three directions of cleavage: A. Cleavages in galena intersect at 90° (cubic cleavage). Note the metallic luster! B. All three cleavages in calcite intersect at either 75° or 105° (rhombohedral cleavage).

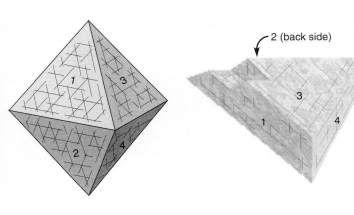

× 1

FIGURE 1.12
Four directions of cleavage (octahedral cleavage) in these samples of fluorite. Numbers indicate different cleavage directions.

× 0.8

FIGURE 1.13
Six directions of cleavage (dodecahedral cleavage) in a sample of sphalerite.

All faces are
cleavage planes

A. Irregular fracture × 1

B. Conchoidal fractures × 1
(See also Figure 3.3.)

FIGURE 1.14
Two types of fracture: A. An irregular or uneven fracture surface on the end of a potassium feldspar cleavage fragment (note the two cleavages at 90° to each other). B. Numerous conchoidal fractures mark the broken surface of this large quartz crystal. Figure 3.3 shows a much larger conchoidal fracture in obsidian.

fractures (Fig. 1.14B). Other kinds of fracture have descriptive names such as fibrous, splintery, or irregular.

In the field you will often have to break samples into pieces to observe cleavages and fractures on fresh surfaces. While it is instructional to hammer some mineral samples yourself, do not break the lab samples without your instructor's approval! Samples cost money and in most cases have already been broken to show characteristic features.

Crystal Form

A **crystal** is a solid, homogeneous, orderly array of atoms and may be nearly any size. Some crystals have smooth, plane faces and regular, geometric shapes (Fig. 1.15); these are what most people think of as crystals. However, a small piece broken from one of these nicely shaped crystals is also a crystal, because the atoms within that small fragment have the same orderly arrangement throughout. When examining minerals, and especially when determining cleavage, you must determine whether you are looking at a single crystal with well-developed crystal faces, a fragment of such a crystal, or a group of small, irregularly shaped intergrown crystals.

Cleavage surfaces may be confused with natural crystal faces; in fact, cleavage planes are parallel to possible (but not always developed) crystal faces. They can be distinguished as follows: (1) Crystal faces are normally smooth, whereas cleavage planes,

× 1

FIGURE 1.15
These crystals of potassium feldspar have smooth, flat faces and regular, geometric shapes.

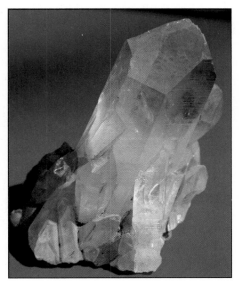

× 1

FIGURE 1.16
Crystal of quartz. Sides form a hexagonal prism that is capped with pyramid-like faces. Note the fine grooves on some crystal faces.

A. × 0.8

B. × 0.8

FIGURE 1.17
Crystals of pyrite in the form of (A) a cube and (B) pyritohedron. Note the fine grooves on the faces of the cubes.

though also smooth, commonly are broken in a step-like fashion (see drawing in Fig. 1.10B or photograph in 1.11A). (2) Some crystal faces have fine grooves or ridges on their surfaces (Figs. 1.16, 1.17A), whereas cleavage planes do not. Similar-looking, very thin, parallel grooves, or *striations,* are seen on plagioclase cleavage surfaces, but

these features persist throughout the mineral and are not surficial, as described below. (3) Finally, unless crystal faces happen to coincide with cleavage planes, the mineral will not break parallel to them.

The arrangement of atoms in a mineral determines the shape of its crystals. Some minerals commonly occur as well-developed crystals, and their crystal forms are diagnostic. A detailed nomenclature has evolved to describe crystal forms, and some

of the common names may be familiar. For example, quartz commonly occurs as *hexagonal* (six-sided) *prisms* with pyramid-like shapes at the top (Fig. 1.16); pyrite occurs as *cubes* (Fig. 1.17A) or *pyritohedrons* (forms with twelve pentagonal faces, Fig. 1.17B); calcite occurs as *rhombohedrons* (six-sided forms that look like cubes squashed by pushing down on one of the corners; Fig. 1.18A) or more complex, twelve-faced forms called *scalenohedrons* (Fig. 1.18B).

A. × 0.5

B. × 0.3

× 1.5

FIGURE 1.18
Crystals of calcite (yellow) in the form of (A) a rhombohedron and (B) a scalenohedron (yellow).

FIGURE 1.19
Calcite reacting to a drop of acid.

Other Properties

Special properties help identify some minerals. These properties may not be distinctive enough in most minerals to help with their identification, or they may be present only in certain minerals.

Reaction with Acid

Some minerals, especially carbonate minerals, react visibly with acid. (Usually, a dilute hydrochloric acid [HCl] is used.) The acid test is especially useful for distinguishing the two carbonate minerals calcite and dolomite. When a drop of dilute hydrochloric acid is placed on calcite, it readily bubbles or effervesces, releasing carbon dioxide (Fig. 1.19). When a drop of acid is put on dolomite, the reaction is much slower unless the dolomite is pow-

dered first; you may even have to look with a hand lens to see the bubbles, or, if the acid is weak, there may not be any. BE CAREFUL when using the acid—even dilute acid can burn your skin or put a hole in your clothing. You need only a small drop of acid to see whether or not the mineral bubbles. When you finish making the test, wash the acid off the mineral immediately. Should you get acid on yourself, wash it off right away, or if you get it on your clothing, rinse it out immediately.

Magnetism

Some minerals are attracted to a hand magnet. To test a mineral for magnetism, just put the magnet and mineral together and see if they are attracted. Magnetite is the only common mineral that is strongly magnetic.

Striations

Plagioclase feldspar can be positively identified and distinguished from potassium feldspar by the presence of *very thin, parallel grooves* called **striations** (Fig. 1.20). The grooves are present on only one of the two sets of cleavages and are best seen with a hand lens. They may not be visible on all parts of a cleavage surface. Before you decide there are no striations, look at all parts of all visible cleavage surfaces, moving the sample around as you look so that light is reflected from these surfaces at different angles.

Until you have seen striations for the first time, you may confuse them with the small, somewhat irregular, differently colored intergrowths or veinlets seen on cleavage faces of some specimens of potassium feldspar (Fig. 1.21). However, these have variable widths, are not strictly parallel, and are not grooves, so they are easily distinguished from striations.

Specific Gravity and Density

The **specific gravity** of a mineral equals its weight divided by the weight of an equal volume of water. The specific gravity of water (at 4°C) equals 1.0, by definition. **Density** is the mass of an object divided by its volume. Specific gravity and density give essentially identical numerical values even though specific gravity, being a ratio of weights, is a unitless number, whereas density is commonly given in units of grams per cubic centimeter. Most of the rockforming minerals (see Chapter 2) have specific gravities of 2.6 to 3.4; the ore minerals (Chapter 2) are usually heavier, with specific gravities of 5 to 8. If you compare similar-sized samples of different minerals with sufficiently different specific gravities, the one with the higher specific gravity will feel heavier. For example, the mineral barite has an unusually high specific gravity (S.G. = 4.5) for a mineral with a nonmetallic luster. When compared with a similar-sized piece of, say, calcite (S.G. = 2.7), the barite will feel much heavier. For most minerals specific gravity is not a particularly noteworthy feature, but for some, such as barite or galena, high specific gravity is distinctive.

Taste, Odor, Feel

Some minerals have a distinctive taste (halite is salt, and tastes like it), some a distinctive odor (the powder of some sulfide minerals, such as sphalerite, a zinc sulfide, smells like rotten eggs), and some a distinctive feel (talc feels slippery).

Tenacity

The **tenacity,** or toughness, of a mineral describes its resistance to being broken. *Brittle* minerals, such as quartz, shatter when broken; *flexible* minerals, like chlorite, can be bent without breaking, but will not resume their original shape when the pressure is released; *elastic* minerals, such as the micas, can be bent without breaking and will spring back to their original position when the pressure is released; *malleable* minerals can be hammered into thin sheets (examples are gold and copper); *sectile* minerals can be cut with a knife (for example, gypsum).

× 0.5

FIGURE 1.20
Striations are visible on the upper surface of this sample of plagioclase.

× 1.5

FIGURE 1.21
The thin veinlets seen in some potassium feldspars should not be confused with striations in plagioclase.

Hands-On Applications

Geologists, like other scientists, must learn to be good observers. Observations provide the basic data on which most science is based. Good observational skills are acquired through training, and experience teaches you which kinds of observations are most important. Observation may mean looking without doing anything else, it may require making some measurements, or it may mean observing the results of a test or experiment. The problems in this first exercise illustrate the kinds of observations needed to identify minerals.

Objectives

If you are assigned all the problems, you should be able to:

1. Determine the color, luster, hardness, and streak of a mineral.
2. Determine whether a mineral has a cleavage or a fracture.
3. Determine the number of cleavages and the angular relations between the cleavages, if a mineral has more than one cleavage.
4. Distinguish between calcite and dolomite using the "acid test."
5. Determine whether a mineral is magnetic.
6. Distinguish between minerals on the basis of specific gravity.
7. Determine whether a sample of feldspar has striations.

Problems

1. Determine whether the luster of each sample provided for you is metallic or nonmetallic. If nonmetallic, indicate whether it is vitreous, pearly, greasy, waxy, resinous, or dull.

2. Determine the streak of each of the samples designated by your instructor.

3. Arrange the designated samples in order of increasing hardness. Determine the Mohs hardness of each.

4. Determine whether each of the samples provided has a cleavage. If it does, determine whether it has 1, 2, 3, 4, or 6 cleavages. If it has 2 or 3 cleavages, estimate whether or not adjoining cleavages intersect at right angles (90°). If only a fracture is present, indicate whether it is a conchoidal fracture.

5. Compare the samples provided by your instructor and list them in order of increasing specific gravity. This is done most easily by holding equal-size samples in each hand and estimating, on the basis of feel, which is the heavier; if the samples are the same size, the heavier one has the higher specific gravity. Alternative methods for determining specific gravity are described in Problem 12.

6. Determine which of the samples provided is attracted to a magnet.

7. Describe the feel of the samples provided.

8. Describe the odor of the mineral supplied by the instructor after rubbing it on a streak plate.

9. Put a small drop of dilute hydrochloric acid (HCl) on the designated samples. Describe what happens and explain why.

10. You have so far seen minerals only under visible light. What about invisible light? It turns out that ultraviolet (UV) light not only darkens skin and causes cataracts, but it also induces some spectacular effects in certain minerals. UV light is relatively safe at longer wavelengths and is found at some dance parties in the form of black lights. Visit the Amethyst Galleries website *http://mineral.galleries.com/* or get to it through *www.mhhe.com/jones6e*—see Preface, click on *Interesting Groupings,* and check out *Fluorescent Minerals* to find out which minerals will look good at a dance club and why.

   a. What is fluorescence?

   b. What is phosphorescence and how does it differ from fluorescence?

   c. Triboluminescence is something you may have observed ripping open Band-Aids™ or crushing Certs™ candies in the dark, but it also occurs with minerals. What is triboluminescence?

   d. Explain whether minerals can always be identified with certainty based on their fluorescence.

   e. You have probably never heard of willemite, but it is a pretty cool mineral. Look it up (click on *By Name,* then *W*) and describe its fluorescence and why specimens from New Jersey excite collectors.

   f. There are two types of UV lights. What are they? Go back to the *Fluorescent Minerals* page and find out if the Franklin, New Jersey, samples would look good at a party equipped with a typical black light.

In Greater Depth

11. People have long wondered what causes crystals to have perfect shapes. If you ever pulled a nice quartz crystal out of some weathered rock, you may have wondered at how something so perfectly ordered came from something so filthy and disordered. The alchemists tried to unravel such mysteries of nature, but their work was unscientific and they could not replicate each other's results. Their data and conclusions were therefore worthless. Nicolaus Steno in 1669 was the first to make objective, scientific measurements related to minerals. He wanted to see if evidence supported the hypothesis that there are underlying 'building blocks' that make up each mineral and determine its basic shape. To do this, he obtained quartz crystals of various sizes and shapes from around the known world. He then measured the angles between adjacent crystal faces and compared them to comparable angles measured on the other samples.

   NOTE: The following exercise can be done with a set of quartz crystals, plus a few amethyst and smoky quartz crystals, or with calcite rhombohedra. The calcite rhombohedra allow a test of whether the angles between cleavage planes vary by size or specimen.

   a. Examine a specimen and select four interfacial angles to compare among all the specimens. Pick some angles that are distinctly different from others on the same sample. Sketch below a cross section and/or profile of a representative sample and number your selected angles. Be sure to label whether each sketch is a top or side view!

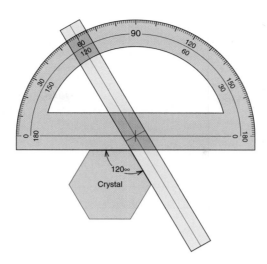

FIGURE 1.22
A simple contact goniometer consists of a protractor and a rotatable arm. The angle between adjacent crystal faces is measured as shown.

b. Using a simple contact goniometer as shown in Figure 1.22, carefully measure the interfacial angles at the four places you indicated. You may want to make several measurements of a single angle to be sure you have a good number. Record these data in the table below.

Angle Number	Sample	Sample	Sample	Sample	Sample	Sample	Sample	Sample	Average
1									
2									
3									
4									

To calculate the average, add the numbers for the same angle on each sample and divide by the total number of samples. For example, if you had four samples on which angle number 1 measured 44, 48, 43, and 45 degrees, the average is $(44 + 48 + 43 + 45)/4 = 45°$.

c. Steno's Law states that comparable interfacial angles are constant from specimen to specimen. Given the scatter in *your* data, do your data support or contradict Steno's Law? You have to decide whether the scatter in your data reflects measurement errors stemming from your simple measurement tool or from real differences that are clearly beyond your measurement error.

d. Compare the clear quartz crystals with amethyst and smoky quartz crystals. Their color is different from clear quartz, but the crystal shapes look similar. Measure your set of reference interfacial angles on the amethyst and smoky quartz to see if there are any significant differences between the colored and clear crystals.

Angle Number	Sample	Sample	Sample	Sample	Average
1					
2					
3					
4					

e. What do your data suggest about the underlying chemical composition and atomic architecture of the colored crystals as compared to clear quartz? If we had sophisticated lab equipment (an X-ray diffractometer), we could test your hypothesis.

12. Science rests on accurate measurements of physical properties. Specific gravity and density are basic properties that can be measured in several ways. Select one or both of the following methods and determine the specific gravity or density of the mineral sample provided by your instructor. Note that all measurements should be made using metric units (in this case, grams, cubic centimeters, and milliliters).

Method 1: Density. 1. Weigh the dry mineral sample in air (m_m). 2. Put 10 to 20 mL of water in a 25 mL graduated cylinder and measure the volume of water (V_w). 3. Carefully submerge the mineral in the water in the cylinder and measure the volume of water and mineral (V_{wm}). 4. Calculate the volume of the mineral (V_m) by subtracting the volume of the cylinder with water only (V_w) from the volume of the cylinder with the mineral (V_{wm}): $V_m = V_{wm} - V_w$. 5. The density is found by dividing the weight (or mass) of the dry mineral by its volume:

$$density = mass/volume, \text{ or } D = m_m/V_m$$

Density typically is given in units of grams/cubic centimeter (gm/cm^3).

For example, say you had a sample that weighed 10.35 gm (m_m). The starting volume in your cylinder (V_w) is 17.3 mL (milliliters). After you put the mineral in the water, the volume rose to 19.6 mL (V_{wm}). The volume of the mineral is $V_{wm} - V_w$ or 19.6 mL − 17.3 mL = 2.3 mL.

$$D = m/V = 10.35 \text{ gm } (19.6 \text{ mL} - 17.3 \text{ mL}) = 10.35 \text{ gm}/2.3 \text{ mL} = 4.5 \text{ gm/mL} = 4.5 \text{ gm/cm}^3$$

Density is typically expressed in gm/cm^3, not gm/mL. Because 1 mL = 1 cm^3, 4.5 gm/mL is the same as 4.5 gm/cm^3. To convert to specific gravity, divide your mineral density by the density of average tap water at 20°C: 0.99823 gm/cm^3, which in this case can be rounded to 1.00 gm/cm^3. Dividing 4.5 gm/cm^3 by 1.00 gm/cm^3 yields 4.5, the specific gravity.

Method 2: Specific Gravity. A second method uses Archimedes's principle, which states that "a body floating or submerged in a fluid is buoyed by a force equal to the weight of the fluid displaced." In other words, if you suspend a mineral in water, its weight will decrease by the weight of water it displaces. This realization caused Archimedes to run naked through the streets shouting "Eureka!" because it allowed him to prove that the King's crown did not have a high enough specific gravity to be pure gold. Using a Jolly balance (or similar equipment), first weigh a sample in air (m_a), then submerge it so it is suspended in water, and weigh it again (m_w). The specific gravity (S.G.) is calculated as:

$$specific\ gravity = S.G. = \frac{weight\ in\ air}{weight\ in\ air - weight\ in\ water} = m_a/(m_a - m_w)$$

For example, if a dry sample of a mineral (pure and without fractures) weighs 122.03 gm, but when submerged in water it weighs 97.65 gm, then

$$S.G. = 122.03 \text{ gm}/(122.03 \text{ gm} - 97.65 \text{ gm}) = 5.01$$

When you have completed your measurements and calculated the specific gravity or density, compare it to the accepted value provided by your instructor. Then assess potential sources of measurement error. For example, consider factors that will affect your weight or volume measurements, such as bubbles attached to mineral grains, impure mineral samples, or difficulties determining precise water-displacement volumes.

CHAPTER 2

Mineral Identification

Materials Needed
- Pencil
- Hand lens
- Set of mineral samples

- Equipment for determination of mineral properties: streak plate, copper penny, glass plate or steel knife, magnet, dilute hydrochloric acid

Introduction

Most of the Earth is made of rocks, and rocks are made of minerals. Just as you had to learn the alphabet before you could read, you need to learn about minerals before you can understand rocks. The first step is to learn how to identify minerals. Chapter 1 shows that minerals have a number of rather easily determined physical properties, and that different minerals have different properties. This chapter explains how to use these properties to identify minerals. First, the text describes the most important (common) minerals found on Earth, and then it prepares you to go from observing the physical properties of an unknown mineral to an accurate identification of that mineral. You will see these same minerals over and over again through the next three labs, so it is worth taking extra care to learn them.

Some of the mineral samples you will see in your laboratory, or that are shown on these pages, may also be considered rocks. They are rocks if they are *aggregates* of crystals of the mineral rather than single crystals. Because large single crystals are very rare or nonexistent for some minerals, the only available samples may be aggregates. In common usage the term *mineral* is used whether referring to a single crystal or aggregates of crystals of the same mineral.

More than 3000 minerals are known to exist, yet only a comparative few are common. The **rock-forming minerals** are the most abundant minerals found in rocks at the Earth's surface. **Minor minerals** may be important constituents of certain rocks, but they are not common in most rocks. **Accessory minerals** are commonly present in rocks, but generally in only small amounts. **Ore minerals** have economic value when concentrated and are mined for the elements they contain. Table 2.1 lists some of the more important minerals within each of these groups; descriptions follow.

Rock-Forming Minerals

Feldspars (K, Na, and Ca Al-Silicates)

As a group, the feldspars are by far the most abundant minerals in the Earth's crust. They are the principal constituents of many kinds of igneous and metamorphic rocks and are abundant in some kinds of sedimentary rocks, as discussed in Chapters 3, 4, and 5. Feldspar is fairly easy to identify due to its two good cleavages that intersect at essentially 90°, a Mohs hardness of 6, and a vitreous luster. There are two principal varieties of feldspar—plagioclase and potassium feldspar.

Plagioclase has a variable chemical composition, ranging from sodium aluminum silicate ($NaAlSi_3O_8$) to calcium aluminum silicate ($CaAl_2Si_2O_8$). **Note: Abbreviations of common chemical elements are inside the back cover.** The color varies as well: sodium-rich varieties tend to be light colored—commonly white, cream to buff, or light gray—whereas calcium-rich varieties tend to be gray to dark gray (Fig. 2.1). However, **striations,** not color, are the distinctive feature of plagioclase (see Fig. 1.20, Chapter 1). Unfortunately, not every plagioclase cleavage shows striations.

Potassium feldspar is a general name that includes several varieties of $KAlSi_3O_8$ (e.g., *orthoclase, microcline, sanidine*). It comes in white, cream, light gray, and even green, but salmon pink is a common and often distinctive color (Fig. 2.2). Some varieties contain semi-parallel veinlets (Fig. 1.21), which, as discussed in Chapter 1, are not to be confused with the striations in plagioclase.

The principal use of feldspar, especially potassium feldspar, is in the production of porcelain and ceramics. Some varieties of Ca-rich plagioclase (called labradorite) show a play of dark blue iridescent colors as the angle of reflected light changes. These are used for ornamental purposes, as is an unusual green variety of microcline known as *amazonite*.

Table 2.1
Some Important Minerals

Rock-Forming Minerals*	Accessory and Minor Minerals	Ore Minerals
Potassium feldspar	Pyrite	Native Copper
Plagioclase feldspar	Magnetite	Graphite
Quartz	Corundum	Sulfur
Hornblende (an amphibole)	Halite	Galena
Augite (a pyroxene)	Fluorite	Sphalerite
Biotite	Apatite	Chalcopyrite
Muscovite	Garnet	Pyrite
Chlorite	Sillimanite	Hematite
Talc	Kyanite	Magnetite
Clays (e.g., kaolinite)	Topaz	Chromite
Olivine	Staurolite	Goethite
Calcite	Epidote	Malachite
Dolomite	Beryl	Azurite
Gypsum	Tourmaline	Barite
Halite	Serpentine	
	Opal	

*In order of decreasing abundance in the Earth's crust.

Quartz (SiO$_2$)

Quartz is one of the most abundant minerals in the crust of the Earth and is common in soils, rocks, and even atmospheric dust. It is hard and tough, so it survives well on the Earth's surface, and it occurs in a wide variety of colors and forms.

Distinctive properties are hardness (H = 7), conchoidal fracture, and vitreous to somewhat greasy luster, but color varies. Quartz comes in several varieties. Well-formed crystals of colorless *rock crystal quartz,* purple *amethyst,* and dark grey *smoky quartz* are common in rock shops, but not in your backyard. Masses of coarsely inter-grown quartz crystals with no well-defined crystal faces occur as white (or iron-stained) *milky quartz* (Fig. 2.3), which is common in many areas, or the rarer pinkish *rose quartz* or grayish *smoky quartz* (Fig. 2.3). Extremely fine-grained varieties

× 0.5

× 0.5

× 0.5

FIGURE 2.1
Various colors of **plagioclase.** Light-color varieties are usually richer in sodium. Darker varieties are richer in calcium.

× 0.5

FIGURE 2.2
Microcline, a variety of **potassium feldspar.**

× 0.5

FIGURE 2.3
Rose **quartz,** common milky quartz, and smoky quartz.

A. Agate × 0.5

B. Chalcedony × 0.5

C. Jasper × 0.5

D. Chert × 0.5

E. Flint × 0.5

F. Petrified wood × 0.5

FIGURE 2.4
Microcrystalline varieties of SiO_2: A. agate; B. chalcedony; C. jasper; D. chert; E. flint; F. petrified wood. Note the color variations in A–F, and the conchoidal fractures in D and E.

of quartz, collectively termed *microcrystalline quartz,* include varicolored banded varieties such as *agate* and *onyx* and more homogeneous varieties including *chalcedony* (brown or gray, and translucent), *jasper* (red), *chert* (gray, white, or pink), and *flint* (dark gray to black) (Fig. 2.4). *Petrified wood* typically is composed of microcrystalline quartz (Fig. 2.4).

Quartz has many uses. It is a major constituent in glass, including the world's best microscope and telescope lenses. Some varieties are used in jewelry (rock crystal, amethyst, tiger's eye [yellowish brown with a wavy or silky sheen], citrine [transparent, yellow to orange-brown]), and both chert and flint were used to make arrowheads, spear points, and other sharp tools. Quartz watches keep good time because a quartz crystal, when hooked to a battery, vibrates at a constant rate. As sand grains or parts of rocks, quartz is also used in concrete, mortar, and building stone.

Amphiboles (Complex Hydrous Na, Ca, Mg, and Fe Al-Silicates)

Hornblende is the most common amphibole. As you can see in Figures 1.9 and 1.10B, hornblende is a black mineral with two directions of good cleavage that meet at roughly 60° or 120°; the cleavage gives it a splintery look, which helps to identify it. Hornblende occurs most commonly as small crystals in combination with other minerals in igneous and metamorphic rocks, but even there, it is distinctive when viewed with a hand lens.

The principal use of amphibole is for a small subset of the group with a fibrous habit that, as with some varieties of serpentine, makes it useful as asbestos. While airborne asbestos fibers can become imbedded in various parts of the lung or abdomen and cause asbestosis, only one variety (the amphibole crocidolite) is known to be especially carcinogenic. Asbestos is a poor conductor

of heat and electricity and can be made into a variety of fabrics. These properties make it useful in many industrial applications, including cements, flooring, friction material, roofing, coatings, packings, gaskets, and protective clothing.

Pyroxenes (Ca, Mg, and Fe Silicates)

Like the amphiboles, most pyroxenes occur as small, crystalline components of rocks, especially certain igneous rocks. The most common pyroxene is *augite,* a greenish-black to black mineral that is easily confused with hornblende. Figure 2.5 shows that augite has two cleavages that intersect at about 90°, and it is not splintery looking like some hornblende. The cleavages may be difficult to see and are not nearly as good as those on hornblende. Another pyroxene, *diopside,* is commonly light green. A few varieties of pyroxene, such as *jadeite,* are prized for their ornamental value.

× 1

FIGURE 2.5
Augite, a variety of pyroxene. Pyroxenes have two good-to-poor cleavages at 90°, but they can be tough to spot.

× 0.5

FIGURE 2.6
Inuit carving of walrus mother and baby in soapstone, a rock composed mostly of **talc,** resting on a platform made of talc. See Figure 1.2F for a more typical lab sample of talc.

Micas (Hydrous K, Mg, and Fe Al-Silicates)

The two most frequently encountered micas, *biotite* and *muscovite,* are both common constituents of igneous and metamorphic rocks. Biotite is black, and muscovite commonly is colorless to pale gray, as you can see in Figures 1.2 and 1.7. Both have one perfect cleavage and a vitreous luster, so they are usually easy to identify. Muscovite is an excellent insulator for both heat and electricity and is used widely in the electrical industry. Big sheets were once used as windows on old wood stoves and in early automobiles (isinglass).

Chlorite (Hydrous Mg and Fe Al-Silicate)

This dark green mineral exhibits one perfect cleavage, but unlike the micas, thin pieces don't spring back to their original position after being bent; pieces are flexible, but not elastic. Most chlorite occurs as small grains in metamorphic rocks, and you may have to use a hand lens to see that it looks like a dark green mica.

Talc (Hydrous Mg Al-Silicate)

As number one on the Mohs hardness scale, talc is soft enough for use even on a baby's soft bottom. Talcum powder is talc. Talc crystals, if large enough, show one perfect cleavage, but in most cases they are too small to see. Instead, the pearly luster (Fig. 1.2F), greasy feel, and softness are distinctive. Talc forms by the alteration (e.g., metamorphism) of preexisting minerals such as olivine, and is the major ingredient in the rock soapstone. You may be familiar with the soapstone carvings of the American Indians (especially peace pipes), Kenyans (decorative plates, bowls, and candle holders), and the Inuits of the Canadian Arctic (Fig. 2.6).

Clays (Hydrous Al-Silicates)

Clay minerals are a significant component of soils, sediment, and sedimentary rocks. But they are not the sort of thing that you would want to put in a display case; in fact, when you dust the display case, you commonly are getting rid of clays! The word *clay* is used in two different ways: it refers to a specific group of minerals and to particles that are extremely small (<0.004 mm [<0.0002 in] diameter). Clay minerals are usually clay-sized particles. Clays are used for all kinds of things, from the manufacture of china and bricks, to stopping leaks in dams, to fillers in paper, paint, medicine (the clay mineral kaolinite is used in Kaopectate™), and chocolate (doesn't that make your mouth water?).

Clays are recognized by their extremely small crystal size, dull luster, softness, and earthy odor. Their colors vary, but the variety most commonly encountered in phys-

× 0.5

FIGURE 2.7
Microscopic crystals of **kaolinite,** a type of clay mineral, comprise 100 percent of this rock. The homogeneous appearance of this sample makes it a good example of a **massive** texture: There are no visible crystal forms or intergrown crystals.

ical geology labs is the white clay *kaolinite* (Fig. 2.7).

Olivine (Mg, Fe Silicate)

Figure 2.8 illustrates a typical sample of olivine. It is yellowish green to green and is granular (sugary), being made of a multitude of small grains (crystals). You have to look at individual crystals to see that it has a conchoidal fracture but no obvious cleavage. And if you check the hardness, make sure that little pieces aren't breaking off, because it is hard (H = 6.5 to 7).

Most olivine occurs in dark-colored igneous rocks. There is a gem variety of olivine (*peridot*), and olivine is used in some heat resistant bricks.

Calcite (CaCO₃) and Dolomite (CaMg(CO₃)₂)

These two carbonates are the major minerals in the common sedimentary rocks limestone and dolostone, and their metamorphosed equivalent, marble. Calcite is also one of the most common minerals to form in veins and open cavities in rocks. When formed in such places, large cleavable pieces and beautiful crystals may develop,

×1

FIGURE 2.8
Olivine. This sample is a mass of irregular, intergrown crystals.

such as those shown in Figure 1.18. Calcite and dolomite are used for many things, among them waste treatment, animal feed, fertilizer, cement, and the manufacture of glass, paper, glazes, enamels, and industrial chemicals.

If you have a piece of calcite that shows cleavage (chances are you will), the three directions of cleavage not at 90° to each other are a giveaway. Add to that the Mohs hardness of 3, and calcite is easy to identify. The easiest way to distinguish calcite from dolomite is to put a drop of dilute hydrochloric acid (HCl) on them to see if they bubble; calcite bubbles strongly, dolomite not so much, or not at all unless powdered.

Gypsum (CaSO₄ · 2H₂O)

Gypsum ranges from a massive, white to gray, carvable variety known as *alabaster* to a white variety with a silky or satiny luster and distinctive fibrous fracture known as *satin spar* to clear crystals or cleavage fragments known as *selenite* (Fig. 2.9). Its hardness—softer than your fingernail—gives it away. The clusters of crystals can be quite beautiful, but they are so fragile that you have to display them in a safe place. Gypsum is used extensively as the basis for plaster and drywall. If you have a clay-rich soil in your garden, you may have used it as a soil conditioner. Gypsum is a common evaporite mineral found as thick layers in sedimentary rocks.

Halite (NaCl)

This is an easy mineral to identify. Just taste it—it's salt. If you don't like the idea of licking where others have licked before, check the other properties. It is generally colorless or white, although impurities may add color. It commonly displays perfect cubic cleavage (see Fig. 1.6) and is soft (H = 2.5). Halite occurs in sedimentary rocks that formed from the evaporation of seawater and commonly is associated with other *evaporite* minerals, such as gypsum. Its principal use is not for a flavoring or for melting ice, but as a source of sodium and chlorine for the chemical industry.

Accessory and Minor Minerals

Fluorite (CaF₂)

Fluorite is a pretty mineral that ranges from colorless to pale yellow, light green, or purple, and frequently occurs as either octahedral cleavage fragments (Fig. 1.12) or as large cubic crystals. It would make a great gemstone except for its softness (Mohs hardness = 4) and ready ability to cleave. It is used primarily in making steel and for the manufacture of hydrofluoric acid. Hydrofluoric acid is one of the few substances that dissolves silicate minerals.

A. × 0.5   B. × 0.5

FIGURE 2.9
Gypsum. A. Satin spar variety. B. Selenite variety showing one perfect and two poor cleavages.

×1

FIGURE 2.10
Serpentine asbestos. Note the fibers.

× 0.8

FIGURE 2.11
Twelve-sided crystals (dodecahedrons) of different varieties and colors of **garnet.**

Serpentine (Hydrous Mg Silicate)

The most familiar type of serpentine is serpentine asbestos (Fig. 2.10), but massive (no crystal forms), green serpentine is most common. The greasy to waxy luster (and feel) of massive serpentine is distinctive; its Mohs hardness (H = 3–5) serves to distinguish it from massive talc (H = 1). Massive-veined serpentine is commonly used as a decorative building stone. More than 95 percent of the asbestos produced in the world is serpentine. Studies have shown that health risks from serpentine asbestos are considerably less than from certain types of amphibole asbestos, although the distinction is rarely recognized by the public.

Garnet (Ca, Mg, Fe, and Al Silicate)

Garnet is a metamorphic mineral that almost always occurs as a perfect crystal (Fig. 2.11). However, some crystals are so large that the sample you see in the lab may be just a piece of a crystal. Mohs hardness (H = 6.5–7.5), vitreous to resinous luster, and crystal form help to identify garnet, but be careful of color. A common color is brownish red, but garnet can be black, brown, yellow, green, or other colors. Garnet is used for abrasives (such as in garnet sandpaper) and gemstones.

×2

FIGURE 2.12
Crystals of **staurolite.** Intergrown crystals on right are *twinned.*

Staurolite (Hydrous Fe Al-Silicate)

Like garnet, staurolite almost always occurs as "well-formed" crystals in metamorphic rocks. Dark brownish crystals (Fig. 2.12) are characteristic of this mineral. Intergrown crystals called *twins* are common and are sometimes sold as "fairy crosses."

Kyanite (Al_2SiO_5)

This metamorphic mineral is light blue and forms blade-shaped crystals (Fig. 2.13). It is used in the manufacture of spark plugs, among other things.

×1

FIGURE 2.13
Kyanite varies from sky blue to blue grey, but frequently shows its *bladed* crystal habit. The mineral surrounding the kyanite is quartz.

Ore Minerals

Hematite (Fe₂O₃)

The red to red-brown streak is the key to identifying hematite, because, as shown in Figure 2.14, it occurs as several strikingly different forms. One form is a reddish brown, soft, earthy material. Another is dull black and hard. A third type, known as *specular hematite,* has a metallic luster and occurs as a mass of small, mica-like scales. However, scratch any hematite, or rub it across a streak plate, and you will see the same red powder. Most hematite occurs in sedimentary rocks (or their metamorphic equivalents) called *iron formations*. It is the most important ore of iron, which is the principal ingredient in steel.

Magnetite (Fe₃O₄)

The only magnetic mineral you are likely to see in this lab is magnetite (Fig. 2.15). But just in case you don't have a magnet, other properties (such as black color and streak, metallic luster on fresh breaks, and moderately high specific gravity) should identify it. Magnetite is a common accessory mineral in igneous and metamorphic rocks. It is an important ore of iron in iron formation and where concentrated by igneous processes.

Limonite (Various Fe oxides)

Limonite is a general name for a mixture of poorly crystalline, hydrated and non-hydrated iron oxides. Although really a rock, it was once considered to be a distinct mineral. The name applies to yellow, yellow-brown, red or black, dull to metallic, soft to hard material. An example is shown in Figure 2.16. It usually forms by oxidation of iron-bearing minerals, so in essence, it is a rust. It has a yellow-brown streak, which serves to distinguish it from hematite, and is a minor ore of iron.

Galena (PbS)

Figures 1.2A and 1.11A show that galena has an obvious metallic luster and three perfect cleavages that intersect at 90°. It also has a very high specific gravity (S.G. = 7.5). Galena is almost the only ore of lead, which is used in batteries and metal products, and occurs most commonly in vein-like deposits.

Sphalerite (ZnS)

This may be the only mineral you will ever see with six directions of cleavage, although you probably will have a difficult time seeing them all (see Fig. 1.13). If the cleavage is hard to see, look for the resinous luster (like tree sap); the yellow, brown, or black color; and the yellow-brown streak with its faint, rotten-egg-like odor of hydrogen sulfide. Sphalerite is the most important ore of zinc and commonly

× 0.5

FIGURE 2.14
Three varieties of **hematite:** specular hematite (left), oolitic hematite, and botryoidal hematite (right).

× 0.5

FIGURE 2.15
The lodestone variety of **magnetite** is a natural magnet strong enough to hold the paper clip. The brownish areas are oxidized ("rusted").

× 0.5

FIGURE 2.16
Limonite. This mixture of iron oxides was once considered a mineral, but is really a rock.

is found with galena. Zinc is used in galvanizing iron or as a metal by itself; it is also used in batteries, paints, dyes, alloys, and dandruff shampoo.

Pyrite (FeS$_2$)

Pyrite is pale brass-yellow, has a metallic luster, and is called *Fool's gold* by those in the know (who, presumably, are not themselves fools). Crystals in the form of cubes and pyritohedrons (see Fig. 1.17) are common, but so are irregular masses that show the characteristic color and uneven fracture. Given these properties, a Mohs hardness of 6 to 6.5 will enable you to distinguish it from other yellow sulfide minerals such as chalcopyrite (H = 3.5–4) or, should the chance arise, gold (H = 2.5–3). It is a common accessory mineral in igneous, sedimentary, and metamorphic rocks and is concentrated in veins or irregular masses in many places. Pyrite is mined along with the more valuable sulfide minerals with which it occurs, and some of it is used in the production of sulfuric acid.

Chalcopyrite (CuFeS$_2$)

This brass-yellow mineral is distinguished from massive pyrite by its slightly more yellow color and its Mohs hardness of 3.5 to 4. It is one of the most important ores of

× 1

FIGURE 2.17
Bauxite with pea-shaped structures. This rock was once considered a mineral.

copper. Copper is used for electrical conductors (such as wire), brass (an alloy with zinc), bronze (an alloy with tin), fittings, sheathings, and other manufactured items. Chalcopyrite occurs in veins and various types of more massive (unlayered, no obvious structure) deposits.

Graphite (C)

Graphite is dark gray, silver gray, or black, and is greasy, soft, and soils your fingers or whatever it touches. Mixed with clay, it forms the "lead" in pencils, and mixed with oil, it is a good lubricant. It has other uses as well, including generator brushes, batteries, and electrodes. It is usually found in metamorphic rocks.

Bauxite (Various Al oxides)

Bauxite is really a rock made of several hydrous Al oxide minerals, but it once was regarded as a mineral. It is characterized by pea-shaped structures (Fig. 2.17); it may be white, gray, or red, and usually is soft, dull, and earthy. It forms by extensive weathering in tropical to subtropical climates. It is important because it is the only ore of aluminum, which is used wherever a light, strong metal is needed.

How to Identify Minerals

Warning: The samples you see in lab may, at first glance, look quite different from the photos in the manual. Nature is expert at providing infinite variation in color and shape, and these are the features most easily captured by camera. In addition, certain photos show museum-quality specimens (for example, pyrite in Fig. 1.17), which are too rare and expensive to use in labs. Thus, you will have to identify your mineral specimens using the physical properties that you learned in Chapter 1 and the tables that follow (Tables 2.2A through F). These tables are designed to allow you to use basic observations to quickly narrow the possibilities to just a few choices. In terms of the scientific method, you will first make some observations and form one or several hypotheses regarding the identity of the mineral. Then you will closely examine the descriptions in the tables to guide

you toward making additional observations that eliminate certain possibilities and help you settle on the best mineral identification. Finally, check to see that the mineral description correctly matches your sample in all important regards.

To streamline the process, we've included a flowchart (Fig. 2.18) to guide you from your unknown mineral to the table that contains its correct identification. For example, suppose you are given a mineral that has the following properties:

Luster	Hardness	Streak
nonmetallic (vitreous)	3.5 to 5	white

Cleavage	Color	Other
four perfect	clear	purplish tinge

According to Figure 2.18, the first determination is luster. Table 2.2A contains minerals with metallic luster; Tables 2.2B through 2.2F contain minerals with nonmetallic luster. Because the unknown mineral has a nonmetallic luster, Table 2.2A is eliminated.

Following the flowchart (Fig. 2.18), the next thing to determine is whether a mineral is harder than glass (H>5.5) or softer than glass (H<5.5). Since our mineral is softer than glass (H is between 3.5 and 5), it must be in Tables 2.2B through 2.2D.

The third observation needed is streak (Fig. 2.18). The white streak eliminates Table 2.2B, leaving only 2.2C and 2.2D.

The last major distinction is based on cleavage. Table 2.2C is for minerals with no cleavage, and 2.2D is for minerals with one or more cleavages. Because our mystery mineral has four good cleavages, it must be in Table 2.2D. Examination of the mineral descriptions in this table shows only one mineral with four cleavages: fluorite.

NOTE: The most useful properties for identifying a given mineral are underlined in the tables. "H" in the descriptions refers to *hardness*, and "S.G." to *specific gravity*. Some minerals appear in more than one table because they may or may not show a particular property in the sample you examine. For example, hematite appears in Table 2.2A, because one variety has a metallic luster, and in Table 2.2B, because another variety does not. Hornblende, augite, and bauxite are listed in two places, because their hardness may be greater or less than 5.5.

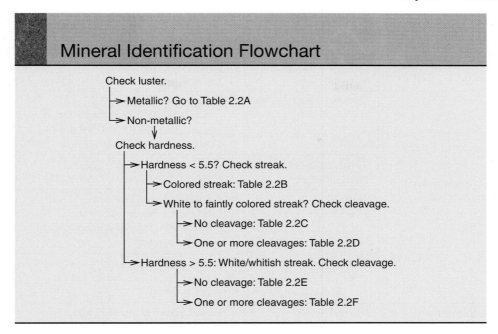

FIGURE 2.18
A flowchart guide to the mineral identification tables.

Since it may be useful in future chapters, Table 2.3 gives the chemical groupings and formulae of the minerals listed in Tables 2.2 A–F.

Insecure About Mineral Identification?

As you work through this lab and the next few that require rock identification, you may feel some insecurity in your mineral identifications. How can you be *sure* that you are correct? Well, all geologists feel this unease at first. With experience, however, confidence and skill grow. This is why every geologist tries to see as many different samples as possible. If you continue in geology, you will also encounter more sophisticated tools used to make more definitive identifications. Geology majors use expensive petrographic microscopes to identify unknown minerals. Graduate students and other researchers use advanced technology to help identify pesky but potentially important unknown minerals. For example, electron microprobes zap samples with electron beams to work out their chemical composition. If more information is needed, an X-ray diffractometer uses X-rays to map out the atomic structure of unknown minerals. Such techniques either provide a definitive mineral name or, with luck, uncover a mineral that is new to science!

Table 2.2A
Metallic Luster or Tarnished (Dulled) Metallic Luster*

Hardness	Streak	Cleavage	Other	Mineral
Less than 5.5	Black, greenish black brownish black, or gray	One (but not always visible)	Gray to black; <u>H = 1; S.G. = 2.5;</u> may have dull luster; greasy feel, <u>soils paper and fingers.</u> **See p. 24.** Also in Table 2.2B.	**Graphite** C
		Three (<u>excellent cubic cleavage</u>)	<u>Silvery gray;</u> H = 2.5; <u>S.G. = 7.5;</u> cubic crystals common. **See p. 23.**	**Galena** PbS
		None	Black to brownish black; H = 5.5; S.G. = 4.6; luster may be almost nonmetallic; <u>brownish-black streak;</u> weakly magnetic.	**Chromite** $FeCr_2O_4$
		None	<u>Golden yellow</u> or greenish yellow; <u>H = 3.5 to 4;</u> S.G. = 4.2; massive. **See p. 24.**	**Chalcopyrite** $CuFeS_2$
	<u>Yellow-brown</u>	None	Dark brown to black; H = 5 to 5.5; S.G. = 4; massive or rounded forms with radiating, fiberlike layers.	**Goethite** FeO(OH)
	Copper-red	None	<u>Copper</u> to brown, may have green coating; H = 2.5 to 3; S.G. = 8.9; <u>malleable, sectile.</u>	**Native Copper** Cu
Greater than 5.5	Greenish black to black	None	<u>Brass-yellow; H = 6 to 6.5;</u> S.G. = 5.0; massive or as crystals (cube, pyritohedron). **See p. 24.**	**Pyrite** FeS_2
	Black	None	Dark gray to black; H = 6; <u>S.G. = 5.2;</u> may have dull luster if surface not fresh; <u>attracted to magnet.</u> **See p. 23.**	**Magnetite** Fe_3O_4
	<u>Red-brown</u>	None	Steel gray to dull red; H = 6; <u>S.G. = 5.0;</u> may be micaceous (tiny flakes) or massive. **See p. 23.** Also listed in Table 2.2B.	**Hematite** Fe_2O_3

<u>Underlined</u> properties are most useful in mineral identification. H = hardness (Mohs). S.G. = specific gravity.

*Like metals, many minerals with a metallic luster oxidize or tarnish as they react with moisture and various compounds in air. This transforms metallic lusters into dull lusters. Geologists in the field break open rocks to expose fresh surfaces (that's why we carry rock hammers!). In this lab you are probably not allowed to smash the samples, so if you have a sample with a dull luster that looks like it could almost be metallic, take a streak and check out the distinctive properties listed in this table.

Table 2.2B
Nonmetallic Luster • Hardness Less than 5.5 • Streak—Colored

Streak	Cleavage	Other	Mineral
Gray-black	One (not always visible)	Gray to black; <u>H = 1; S.G. = 2.5;</u> may have metallic luster; greasy feel; soils paper and fingers. **See p. 24.** Also listed in Table 2.2A.	**Graphite** C
Red-brown	None	<u>Red to reddish brown;</u> H = 1.5 to 5.5; S.G. = 5.0; dull luster; earthy or oolitic (containing spherical structures having diameters of 0.25 to 2 mm) masses. **See p. 23.** Also listed in Table 2.2A.	**Hematite** Fe_2O_3
Yellow-brown	None	<u>Yellow-brown, orange-brown</u> to dark brown; H = 1.5 to 5.5; S.G. = 3.6 to 4.0; dull luster, earthy, may be a rustlike coating. Not a true mineral. Limonite is a general name for several rustlike, hydrous iron oxides. **See p. 23.**	**Limonite** Various hydrous iron oxides
Yellow	None	<u>Yellow;</u> H = 1.5 to 2.5; S.G. = 2.1; resinous or vitreous luster.	**Sulfur** S
	<u>Six</u>—may be difficult to count	Light yellow, yellowish brown, black; H = 3.5 to 4; S.G. = 4.0; <u>resinous or vitreous luster;</u> faint odor of hydrogen sulfide (rotten eggs). **See p. 23.**	**Sphalerite** ZnS
Blue	None	<u>Azure blue;</u> H = 3.5 to 4; S.G. = 3.8; occurs as coatings, masses, or tiny crystals, <u>commonly with malachite,</u> and commonly as an accessory mineral.	**Azurite** $Cu_3(CO_3)_2(OH)_2$
Green	None	<u>Bright green;</u> H = 3.5 to 4; S.G. = 4.0; occurs as coatings, masses, or tiny crystals, <u>commonly with azurite,</u> and commonly as an accessory mineral.	**Malachite** $Cu_2CO_3(OH)_2$

<u>Underlined</u> properties are most useful in mineral identification. H = hardness (Mohs). S.G. = specific gravity.

Table 2.2C
Nonmetallic Luster • Hardness Less than 5.5
• Streak—White, Whitish, or Faintly Colored • No Apparent Cleavage

Cleavage	Other	Mineral
None or Indistinct	White, may have brownish tint; <u>H = 1 to 2.5;</u> S.G. = 2.6; dull luster; greasy feel, earthy odor, powdery. **See p. 20.**	**Kaolinite** $Al_2Si_2O_5(OH)_4$
	White, gray, or apple green; pearly luster; H = 1 (or 2 in some impure varieties); S.G. = 2.7 to 2.8; <u>greasy feel;</u> may have light gray streak. **See p. 20.** Also listed in Table 2.2D.	**Talc** $Mg_3Si_4O_{10}(OH)_2$
	White; <u>H = 2;</u> S.G. = 2.3; Satiny or silky luster and fibrous fracture indicates *satin spar* variety of gypsum. An opaque, white to gray, massive variety is *alabaster*. **See p. 21.** Also in Table 2.2D.	**Gypsum** $CaSO_4 \cdot 2H_2O$
	Reddish brown to brown color; H = 1 to 6; S.G. = 2.0 to 2.5; dull to earthy luster; massive or with <u>pea-shaped structures;</u> may scratch glass with difficulty. Not a true mineral, but a rock consisting of several similar minerals. May have faint, red-brown streak. **See p. 24.** Also listed in Table 2.2E.	**Bauxite** Mixture of hydrous aluminum oxides
	Multicolored green, gray, black; H = 3 to 5; S.G. = 2.5 to 2.6; dull to <u>greasy luster;</u> <u>slight greasy feel;</u> massive to fibrous (*asbestos*). **See p. 22.**	**Serpentine** $Mg_3Si_2O_5(OH)_4$
	<u>Light green to medium green,</u> brown, yellow; vitreous luster; <u>H = 5;</u> S.G. = 3.2; six-sided crystals common; may show one poor cleavage.	**Apatite** $Ca_5(PO_4)_3(F,Cl,OH)$
	Buff, gray, white, pinkish; H = 3.5 to 4; S.G. = 2.8 to 2.9; small, rhombohedral crystals or massive; three cleavages, not at 90°, may be indistinct in massive varieties; <u>reacts slowly or not at all with dilute hydrochloric acid unless powdered.</u> **See p. 21.** Also listed in Table 2.2D.	**Dolomite** $CaMg(CO_3)_2$

<u>Underlined</u> properties are most useful in mineral identification. H = hardness (Mohs). S.G. = specific gravity.

Table 2.2D
Nonmetallic Luster • Hardness Less than 5.5
• Streak—White, Whitish, or Faintly Colored • One or More Cleavages

Cleavage	Other	Mineral
One	Light apple green, gray, white; pearly luster; H = 1; S.G. = 2.7 to 2.8; cleavage not evident in finely crystalline, massive varieties; may have light gray streak; greasy feel. **See p. 20.** Also listed in Table 2.2C.	**Talc** $Mg_3Si_4O_{10}(OH)_2$
	Green to blackish green; dull to vitreous or pearly luster; H = 2 to 2.5; S.G. = 2.6 to 3.3; may have faint green-yellow streak; cleavage flakes are flexible but not elastic; finely crystalline aggregates common. **See p. 20.**	**Chlorite** Hydrous Mg-Fe-Al silicate
	Black to brownish black; vitreous luster; H = 2.5 to 3.0; S.G. = 2.8 to 3.2; may have faint brown-gray streak; individual crystals commonly are small and cleavage surfaces are wavy; perfect cleavage; transparent, flexible and elastic in thin sheets. **See p. 20.**	**Biotite** Hydrous K-Mg-Fe-Al silicate
	Colorless, silvery white, brownish silvery white; vitreous luster; H = 2.0 to 2.5; S.G. = 2.8 to 2.9; perfect cleavage; transparent, flexible, and elastic in thin sheets. **See p. 20.**	**Muscovite** Hydrous K-Al silicate
	Clear, white, light gray; H = 2; S.G. = 2.3; vitreous to pearly luster; brittle sheets; one perfect cleavage, two poor cleavages indicate the *selenite* variety of gypsum; *alabaster* is massive, *satin spar* is fibrous. **See p. 21.** Also listed in Table 2.2C.	**Gypsum** $CaSO_4 \cdot 2H_2O$
Two	Black; H = 5 to 6; S.G. = 3.0 to 3.4; vitreous luster; two perfect cleavages meet at 124° and 56°; cleavage faces stepped rather than smooth; splintery appearance; may have greenish black to black streak. An amphibole. **See p. 19.** Also listed in Table 2.2F.	**Hornblende** Hydrous Na-Ca-Mg-Fe-Al silicate
	Black to dark green; H = 5 to 6; S.G. = 3.2 to 3.4; vitreous to dull luster; two imperfect cleavages meet at nearly 90°. Tends to look 'blocky,' not splintery. A pyroxene. Another pyroxene, *diopside,* is similar but is light grayish green. **See p. 19.** Also listed in Table 2.2F.	**Augite** Ca-Mg-Fe silicate
Three	Clear to gray to red; H = 2.5; S.G. = 2.2; three perfect cleavages meet at 90° (cleavage surfaces may dull and partially dissolve with prolonged exposure); salty taste. **See p. 21.**	**Halite** NaCl
	Clear, white, light gray; H = 2; S.G. = 2.3; vitreous to pearly luster; brittle sheets; one perfect cleavage, two poor cleavages may be evident in clear pieces. **See p. 21.** Also in Table 2.2C.	**Gypsum** $CaSO_4 \cdot 2H_2O$
	Colorless or white; H = 3 to 3.5; S.G. = 4.5—heavy for a nonmetallic; tabular crystals, roselike array of crystals, or massive; three cleavages meet at 90°.	**Barite** $BaSO_4$
	Clear, white, less commonly other colors; vitreous luster; H = 3; S.G. = 2.7; three perfect cleavages form rhombic cleavage fragments; double image seen through clear pieces; reacts strongly with dilute hydrochloric acid. **See p. 21.**	**Calcite** $CaCO_3$
	Buff, gray, white, pinkish; H = 3.5 to 4; S.G. = 2.8 to 2.9; small, rhombohedral crystals or massive; three cleavages, not at 90°, may be indistinct; reacts slowly or not at all with dilute hydrochloric acid unless powdered. **See p. 21.** Also listed in Table 2.2C.	**Dolomite** $CaMg(CO_3)_2$
Four	Purple, green, blue, yellow, clear; H = 4; S.G. = 3.2; vitreous luster; perfect cleavage up to four directions may yield octahedral cleavage fragments; may occur as cubic crystals. **See p. 21.**	**Fluorite** CaF_2

Underlined properties are most useful in mineral identification. H = hardness (Mohs). S.G. = specific gravity.

Table 2.2E
Nonmetallic Luster • Hardness Greater than 5.5
• Streak—White or Whitish if Softer than Streak Plate • No Cleavage

Cleavage	Other	Mineral
None	Brown, pink, blue, gray; <u>H = 9</u>; S.G. = 4.0; <u>six-sided prismatic crystals</u>; *ruby* (red) and *sapphire* (commonly blue) are gem varieties.	**Corundum** Al_2O_3
	<u>Black</u>, pink, blue, green, brown; vitreous luster; H = 7 to 7.5; S.G. = 3.0 to 3.3; <u>slender crystals with triangular cross sections and striated sides.</u>	**Tourmaline** Complex hydrous silicate
	Reddish brown, yellowish tan; vitreous to resinous luster; <u>H = 6.5 to 7.5</u>; S.G. = 3.6 to 4.3; <u>twelve-sided crystals</u> common. Broken surfaces may resemble cleavage in some samples. **See p. 22.**	**Garnet** Ca-Mg-Fe-Al silicate
	<u>Red-brown to brownish black</u>; vitreous, resinous, or dull luster; H = 7 to 7.5; S.G. = 3.7; <u>prismatic and X- or cross-shaped crystals. See p. 22.</u>	**Staurolite** Hydrous Fe-Al silicate
	Coarsely crystalline varieties: clear, milky, white, purple, smokey, pink; transparent to translucent; vitreous luster; <u>H = 7</u>; S.G. = 2.7; <u>conchoidal fracture</u>; usually massive; sometimes occurs as six-sided crystals; milky = *milky quartz,* purple = *amethyst,* smoky = *smoky quartz,* pink = *rose quartz.* Microcrystalline varieties: *chert* (gray, dull luster), *flint* (black, dull luster), *chalcedony* (brown to gray, translucent, waxy luster), *agate* and *onyx* (varicolored bands, vitreous luster). **See pp. 18 and 19.**	**Quartz** SiO_2
	Colorless, white, or pale shades of yellow, green, red, or blue; may show <u>play of colors</u>; vitreous to resinous luster; H = 5 to 6; S.G. = 2.0 to 2.3; <u>rounded forms common,</u> but also massive; conchoidal fracture. A mineraloid.	**Opal** $SiO_2 \cdot nH_2O$
	<u>Olive green to yellow green</u>; vitreous to dull luster; H = 6½ to 7 but difficult to test because <u>granular</u>; S.G. = 3.3 to 4.4. **See p. 20.**	**Olivine** $(Mg,Fe)_2SiO_4$
	Reddish brown to brown color; H = 1 to 6; S.G. = 2.0 to 2.5; dull to earthy luster; massive or with <u>pea-shaped structures</u>; may scratch glass with difficulty. Not a true mineral, but a rock consisting of several similar minerals. **See p. 24.** Also listed in Table 2.2C.	**Bauxite** Mixture of hydrous aluminum oxides

<u>Underlined</u> properties are most useful in mineral identification. H = hardness (Mohs). S.G. = specific gravity.

Table 2.2F
Nonmetallic Luster • Hardness Greater than 5.5 • Streak— White or Whitish if Softer than Streak Plate • One or Two Cleavages

Cleavage	Other	Mineral
One	Colorless, yellow, brown, pink, bluish; <u>H = 8;</u> S.G. = 3.4 to 3.6; vitreous luster; <u>elongate crystal prisms</u> with pointed ends and striated side faces.	**Topaz** $Al_2SiO_4(OH,F)_2$
	<u>Bluish green,</u> yellow, white, pink; <u>H = 7.5 to 8;</u> S.G. = 2.7 to 2.8; elongate, six-sided crystal prisms with flat ends common.	**Beryl** $Be_3Al_2(Si_6O_{18})$
	<u>Light blue to greenish blue;</u> vitreous luster; H = 5 parallel to long direction of crystal, seven across crystal; <u>blade-shaped crystals;</u> one cleavage. **See p. 22.**	**Kyanite** Al_2SiO_5
	White, pale green, brown; H = 6 to 7; S.G. = 3.2; <u>long, slender crystals, commonly as groups of parallel crystals.</u>	**Sillimanite** Al_2SiO_5
Two	<u>Salmon-pink, white,</u> gray, green; vitreous luster; H = 6; S.G. = 2.5 to 2.6; <u>two cleavage directions meet at nearly right angles; no striations.</u> **See p. 17.**	**Potassium feldspar** $KAlSi_3O_8$
	<u>White to dark gray,</u> sometimes buff; vitreous luster; H = 6; S.G. = 2.6 to 2.8; <u>two cleavage directions meet at nearly right angles;</u> some cleavage faces have perfectly <u>straight, parallel striations,</u> which show up in reflected light. **See p. 17.**	**Plagioclase feldspar** $NaAlSi_3O_8$ to $CaAl_2Si_2O_8$
	<u>Pistachio green,</u> yellowish green; <u>H = 6 to 7;</u> S.G. = 3.3 to 3.5; elongate crystals or finely crystalline masses.	**Epidote** Hydrous Ca-Fe-Al silicate
	<u>Black;</u> H = 5 to 6; S.G. = 3.0 to 3.4; vitreous luster; <u>two perfect cleavages meet at 124° and 56°;</u> cleavage faces stepped rather than smooth, giving splintery appearance; may have faint greenish-gray streak. An amphibole. **See p. 19.** Also listed in Table 2.2D.	**Hornblende** Hydrous Na-Ca-Mg-Fe-Al silicate
	<u>Black to dark green;</u> H = 5 to 6; S.G. = 3.2 to 3.4; vitreous to dull luster; <u>two imperfect cleavages meet at nearly 90°.</u> Tends to look 'blocky,' not splintery. A pyroxene. Another pyroxene, *diopside,* is similar but is light gray to <u>light grayish green.</u> **See p. 19.** Also listed in Table 2.2D.	**Augite** Ca-Mg-Fe silicate

<u>Underlined</u> properties are most useful in mineral identification. H = hardness (Mohs). S.G. = specific gravity.

Table 2.3
Chemical Grouping and Formulae of the Minerals in Tables 2.2A–F

Chemical Group	Mineral Name	Chemical Formula
Elements	Copper	Cu
	Sulfur	S
	Graphite	C
Sulfides	Galena	PbS
	Sphalerite	ZnS
	Chalcopyrite	$CuFeS_2$
	Pyrite	FeS_2

	Corundum	Al_2O_3
Oxides and hydroxides	Hematite	Fe_2O_3
	Magnetite	Fe_3O_4
	Chromite	$FeCrO_4$
	Goethite	$FeO(OH)$
Halides	Halite	$NaCl$
	Fluorite	CaF_2
Carbonates	Calcite	$CaCO_3$
	Dolomite	$CaMg(CO_3)_2$
	Malachite	$Cu_2CO_3(OH)_2$
	Azurite	$Cu_3(CO_3)_2(OH)_2$
Sulfates	Barite	$BaSO_4$
	Gypsum	$CaSO_4 \cdot 2H_2O$
Phosphates	Apatite	$Ca_5(PO_4)_3(F,C,OH)$
Silicates[1]		
Isolated SiO$_2$ tetrahedra	Olivine	$(Mg,Fe)_2SiO_4$
	Garnet	$(Mg,Fe,Ca)_3(Al,Fe)_2(SiO_4)_3$
	Kyanite and sillimanite	Al_2SiO_5
	Topaz	$Al_2SiO_4(OH,F)_2$
	Staurolite	$Fe_2Al_9O_6(SiO_4)_4(O,OH)_2$
Double SiO$_2$ tetrahedra	Epidote	$Ca_2(Fe,Al)Al_2O(SiO_4)(Si_2O_7)(OH)$
Rings of SiO$_2$ tetrahedra	Tourmaline	$(Na,Ca)(Li,Mg,Al)(Al,Fe)(BO_3)_2(Si_6O_{18})(OH)_4$
	Beryl	$Be_3Al_2(Si_6O_{18})$
Single chains of SiO$_2$ tetrahedra (Pyroxene group)	Augite	$Ca(Mg,Fe)(SiO_3)_2$
Double chains of SiO$_2$ tetrahedra (Amphibole group)	Hornblende	$(Ca,Na)_{2-3}(Mg,Fe,Al)_5Si_6(Si,Al)_2O_{22}(OH)_2$
Sheet of SiO$_2$ tetrahedra	Serpentine	$Mg_3Si_2O_5(OH)_4$
	Kaolinite	$Al_2Si_2O_5(OH)_4$
	Talc	$Mg_3Si_4O_{10}(OH)_2$
	Muscovite	$KAl(AlSi_3O_{10})(OH)_2$
	Biotite	$K_2(Mg,Fe)_3(AlSi_3O_{10})(OH)_2$
	Chlorite	$(Mg,Fe)_3(Si,Al)_4O_{10}(OH)_2 \cdot (Mg,Fe)_3(OH)_6$
Framework of SiO$_2$ tetrahedra	Quartz	SiO_2
	Opal	$SiO_2 \cdot nH_2O$
	K-feldspar	$KAlSi_3O_8$
	Plagioclase feldspar	$NaAlSi_3O_8-CaAl_2Si_2O_8$

[1]The silicates are grouped according to the way in which the silica (SiO_2) tetrahedra are arranged in their crystal structures. Many of the silicates have variable chemical compositions, and the formulae given in the table may be representative only.

Hands-On Applications

Just as bacteriologists need to be able to identify different kinds of bacteria, geologists need to know what the most common minerals look like and how to identify ones they don't already know. In Chapter 1, you learned how to determine some of the physical properties of minerals. The present chapter illustrates how the basic physical properties, arranged in the form of an identification key or determinative tables, provide a rational, straightforward method for mineral identification. Identification keys or tables, like the ones used for minerals, are a means of organizing large amounts of information. They can be used to identify trees, flowers, birds, animals, or any group of things that can be organized on the basis of similarities and dissimilarities. Organization of knowledge is one of the ways in which science progresses; it would be terribly inefficient if no general procedures were available to identify things like minerals.

Objectives

If you are assigned all the problems, you should be able to identify those minerals specified by your instructor.

Problems

1. Determine the physical properties of the mineral samples provided by your instructor and identify each sample using the mineral identification tables. You may find it helpful to use the tear-out Mineral Identification Worksheets, which are on the last pages of this chapter.

2. Distinguish among similar minerals:
   a. Both pyrite and chalcopyrite have a metallic luster, a greenish-black streak, and no cleavage; they are heavy and can be very difficult to distinguish on the basis of color. If no crystals are visible, how would you tell them apart?

   b. Magnetite and hematite can both be dark gray to black and look similar. How could you tell them apart if you did not have a magnet?

   c. Hornblende and augite commonly are black, have the same hardness, and exhibit two directions of cleavage. They are difficult to distinguish when they occur as small crystals in a rock. What would you look for to tell them apart?

   d. Calcite, halite, and fluorite all have perfect cleavages, and they can all be the same color. How would you distinguish among them?

   e. Talc and serpentine can be the same color, have similar greasy to pearly lusters, and both can have a greasy feel. How do they differ?

   f. If chlorite and biotite occur as small crystals in rocks, it may be difficult to tell them apart. What properties might be helpful?

   g. What single property is most useful for distinguishing between potassium feldspar and plagioclase?

3. Go to *http://mineral.galleries.com* (or link to it through *www.mhhe.com/jones6e*—see Preface) and link to *Birthstones* through *Interesting Groupings.* Look up your birthstone.

    a. What is your birthstone?

    b. What is its chemical formula?

    c. What is its color? Luster? Hardness? Cleavage? Streak?

    d. With what other minerals does it commonly occur?

    e. Where are its notable occurrences?

    f. Is it a variety of another type of mineral or a member of a mineral group?

    Other websites of interest include *www.theimage.com,* which contains pictures of, and information on, many minerals; and *www.gemstone.org/,* a site that has pictures and lots of information about gems.

In Greater Depth

4. Acid mine drainage is a problem that affects many rivers and streams in the coal mining regions of the Appalachians (especially Pennsylvania and West Virginia) and in watersheds surrounding coal and metal mines across the United States, Canada, and the rest of the world. Do a World Wide Web search using the phrase "acid mine drainage" to answer the following questions. *www.google.com* is a good search engine. If you are interested in acid mine drainage in a particular state or country (and you are pretty sure that it has coal or metal ore mines in it), add the name to your search.

    You will get a lot of hits from an open search. Read the questions and pick a site or two that seems informative and authoritative.

    a. What is the name and web address of the site or sites you selected?

    b. Indicate whether it was set up by a government organization, an independent environmental organization, an environmental company, a university research group, or another type of individual or organization.

    c. What mineral or minerals are causing the acid mine drainage?

    d. How does acid mine drainage affect the chemistry of the streams it flows into?

    e. What is one visually obvious symptom of acid mine drainage?

    f. The question of how to remediate (fix) the problems associated with acid mine drainage is interesting, but takes us too far beyond the scope of this lab. Instead, given the owner of the website and the way the information was presented, do you think the information was wholly objective and reliable, or do you think it was slanted by a political or economic agenda?

Mineral Identification Worksheet

Specimen Number	Luster	Hardness	Streak	Cleavage	Other Properties	Mineral Name	Chemical Grouping (Table 2.3)*

*This is useful information to remember if you want to continue in a geology or environmental program.

Mineral Identification Worksheet

Specimen Number	Luster	Hardness	Streak	Cleavage	Other Properties	Mineral Name	Chemical Grouping (Table 2.3)*

*This is useful information to remember if you want to continue in a geology or environmental program.

Part II

Rocks Beach pebbles, Lake MacDonald, Glacier National Park, MT × 2

Igneous Rocks

Materials Needed
- Pencil
- Hand lens
- Calculator
- Samples of igneous rocks

Introduction

Most **rocks** are aggregates of crystals or grains of one or more minerals. The individual mineral particles in rocks are generally small, with average dimensions less than 1 centimeter (cm). Even so, the minerals are identifiable. They have the same physical properties as the larger mineral specimens you have already studied.

Deciphering the Earth's history begins with rocks because they bear testimony to their origin and subsequent history. Once you have completed the next three chapters, you will be able to use easily observable details to infer at least some of the history reflected in the rocks around you.

Rocks are classified as igneous, sedimentary, and metamorphic based on how they formed: **igneous rocks** solidified from melted rock, **sedimentary rocks** form from material (sediment) deposited on the Earth's surface, and **metamorphic rocks** form when preexisting rocks recrystallized, as solids, under high temperature and pressure beneath the Earth's surface. These basic rock types are linked by a set of processes that routinely converts igneous rocks to sedimentary and metamorphic rocks, sedimentary rocks to igneous and metamorphic rocks, and metamorphic rocks to sedimentary and igneous rocks. This global rock recycling program is called the **rock cycle.** In this chapter you will learn to interpret and identify common igneous rocks.

Igneous Rocks

The most spectacular way in which igneous rocks form is by volcanic eruption. This happens when molten rock, called **magma,** rises to the surface. The magma then either simply flows onto the surface as **lava** or violently explodes at the surface due to rapid bubble formation within. The forming bubbles increase the volume of the magma so dramatically that it shoots up and out of the volcano. Igneous rocks that cool on the surface are called **extrusive rocks.** Extrusive rocks cool quickly because air, rain, and snow rapidly remove a lot of heat.

Most rising magmas never make it to the surface. They stall deep in the crust and slowly cool. Igneous rocks that cool beneath the surface are termed **intrusive.** The trapped magma cools slowly because the surrounding rocks form an insulating jacket that helps the magma to retain its heat. Because temperature increases with depth in the Earth, the surrounding rocks are themselves warm and this, too, promotes slow cooling.

Igneous rocks are widely studied for many reasons. First, many people live close enough to volcanoes to be killed by explosive eruptions or huge landslides triggered by eruptions. Geologists study the ancient deposits of individual volcanoes to understand their likely future eruptive style. Second, much of the Earth's crust, both continental and oceanic, is made of igneous rocks; thus, we look to igneous rocks to understand how our crust formed and why our planet is so different from the other rocky planets. Third, igneous rocks commonly form at tectonic plate boundaries; hence, the study of ancient igneous rocks tells us a great deal about the history of the Earth. For example, igneous rocks tell us that about one billion years ago North America partially split along a zone stretching from Lake Superior to northeastern Kansas. Finally, hot magmas drive the circulation of a lot of hot water. These hot fluids pick up a number of important metals and can deposit them to create metal ore deposits. These deposits of gold, copper, silver, and a host of other rare metals form the basis of our modern technological economy.

Although all igneous rocks by definition cool from a liquid magma, they form under such diverse conditions that they differ widely in texture, mineral composition, and appearance. Let's cover these basic differences first.

Texture

Texture refers to the size, shape, and arrangement of the crystals or grains composing the rock. The texture of a rock reflects its composition and the way in which it formed.

With the exception of some volcanic rocks, all igneous rocks have a **crystalline** texture (Fig. 3.1), in which the various mineral crystals are interlocked with one another. This texture develops when crystals grow together as a magma solidifies to form solid rock.

We illustrate rock textures using not only hand specimen photographs but also microscopic views using photomicrographs (Fig. 3.1). Photomicrographs—in this case, photographs of very thin (0.03 mm), transparent slices of rock called **thin sections**—are taken through a microscope equipped with polarizing filters. These filters produce distinctive colors and other optical properties that enable individual minerals in a rock to be confidently identified. The filtered colors are *not* natural, but textures are clearly revealed.

The rate at which a magma cools has the greatest effect on the sizes of the crystals in an igneous rock. In general, the more slowly a magma cools, the larger the mineral crystals will be, because slow cooling provides more time for the chemical constituents to migrate to the growing mineral. As you will see shortly, water and gas content can also profoundly affect rock textures.

Textures Based on Crystal Size

Coarse-grained (phaneritic): An igneous texture in which nearly all crystals are large enough to be seen without a hand lens (Fig. 3.1). For convenience, this is taken to be crystals greater than 1 mm (0.04 in). This texture suggests slow cooling of the magma.

Fine-grained (aphanitic): An igneous texture in which most crystals are too small to recognize with the unaided eye (< 1 mm) (Fig. 3.2). This texture suggests fast cooling of the magma or lava.

Glassy: Describes igneous rocks that are made of glass and lack crystals (Fig. 3.3). Like fine-grained textures, glassy textures suggest rapid cooling. In addition, glassy rocks are often high in silica (SiO_2). The combination of fast

Quartz

Plagioclase

Biotite

Potassium feldspar

Photomicrograph × 10

× 1

FIGURE 3.1

A coarse-grained crystalline texture clearly shows intergrown mineral crystals. Potassium feldspar, quartz, biotite, and plagioclase are visible in this sample of **granite.** The inset photomicrograph emphasizes how tightly intergrown the crystals are. **The colors in this and most subsequent photomicrographs are not natural; they have been changed by polarizing filters placed on the microscope.**

Photomicrograph × 10

Basalt × 1

FIGURE 3.2

A fine-grained, or aphanitic, texture is typical of **basalt.** Its crystalline texture is visible under a microscope (inset photomicrograph).

cooling and high silica makes a lava solidify so fast that its atoms do not have time to form mineral crystals. Instead, the disordered atoms make glass. Glassy rocks may contain some crystals.

Pegmatitic: Describes igneous rocks with exceptionally large crystals (> 3 cm [1.2 in]). Such rocks are called *pegmatites* (see Fig. 3.7). Large crystals result not only from slow cooling but

also from extra water in the magma. Water promotes rapid crystal growth by speeding migration of chemical elements to the growing crystals.

Porphyritic: An igneous texture in which crystals of two different sizes are present in the same rock (Fig. 3.4). The larger crystals, called **phenocrysts,** are surrounded by many smaller crystals, collectively called the **groundmass.** The

Obsidian × 1

FIGURE 3.3

Obsidian is easy to identify because of its glassy texture and vitreous (glassy) luster. The excellent conchoidal fracture produces edges that are sharper than any made from metals. This characteristic made obsidian a favorite among Native Americans for arrows and cutting tools and, more recently, it has even been used for very delicate eye surgery. It is most commonly black (due to inclusions of tiny crystals of magnetite) but may be reddish (inclusions of hematite). *Snowflake obsidian* has white spots that formed where the glass devitrified (crystallized slightly), usually to a variety of silica (SiO_2).

× 1

FIGURE 3.4

Porphyritic texture in hand specimen of **basalt.** Large light-colored crystals (plagioclase phenocrysts) are set in a fine-grained, dark-colored groundmass.

groundmass can be either coarse grained or fine grained, but the phenocrysts are typically coarse grained. Although both intrusive and extrusive rocks can be por-

phyritic, it is more typical of extrusive rocks. A porphyritic texture suggests an initial period of slow cooling followed by faster cooling. For example, phenocrysts may grow in a slowly cooling magma beneath a volcano. If the volcano then erupts, rapid cooling of the remaining liquid would produce a fine-grained groundmass.

Additional Extrusive Rock Textures

Extrusive rocks can develop a range of additional textures if their magmas have high water and gas concentrations. A rising magma experiences a dramatic drop in pressure as it reaches the surface. This release of pressure allows bubbles to form in the magma (or lava), much as bubbles form in a carbonated beverage just after the bottle is opened.

Vesicular: Bubbles that form in lava and neither pop nor escape can be frozen into the lava. The cavities left behind are called **vesicles;** the texture is called **vesicular.** Vesicles range from a few millimeters to several centimeters in diameter. Some rocks contain just a few vesicles; others are more vesicle than rock. Some water and gas-rich magmas form so many bubbles that the lavas froth out onto the surface, solidify very rapidly, and form a rock with lots of millimeter-sized vesicles. *Pumice* and *scoria,* described under Fig. 3.16, are examples.

Pyroclastic: Explosive volcanic eruptions expel a lot of gas, glassy volcanic ash, lava, and fragments of rocks and minerals. All of this material comes to rest on the ground following either a free fall through the air or a powerful gas-charged surface flow down the side of the volcano. Either way, the resulting deposit contains a lot of shattered volcanic glass, rock, and mineral fragments, collectively termed **pyroclasts,** that define a **pyroclastic texture.** An extrusive igneous rock with a pyroclastic texture, like that shown in Figure 3.5, is called a **tuff.** Air-fall tuffs frequently show layering, which is otherwise unusual in igneous rocks. Thin pyroclastic deposits formed from flows cool rapidly and, because of their mode of

Rhyolite tuff × 1

FIGURE 3.5

Pyroclastic texture in hand specimen of **rhyolite tuff.**

formation, produce unusually soft and lightweight rocks with rather open, porous textures. Thick, hot pyroclastic deposits can retain enough heat for the glassy fragments to fuse (weld) together to form a dense, hard, compacted tuff known as a *welded tuff.*

Mineral Composition

Relatively few minerals make up most igneous rocks. To correctly identify the rock, you must identify the major minerals. This is fairly easy for coarse-grained rocks, more difficult (but not necessarily impossible) for fine-grained rocks, and impossible for glassy rocks (unless phenocrysts are present). Table 3.1 and the following may be helpful:

1. Most dark-colored igneous rocks are rich in calcium plagioclase and ferromagnesian (iron-magnesium) minerals such as pyroxene or olivine. The word **mafic** refers to such rocks. **Ultramafic** igneous rocks are composed entirely of ferromagnesian minerals.

2. Light-colored or **felsic** igneous rocks commonly contain potassium feldspar, sodium plagioclase, and quartz, and only minor amounts of other minerals.

3. **Intermediate** igneous rocks are neither dark nor light and generally contain light-colored minerals (feldspars, minor quartz) and dark minerals such as hornblende or biotite.

4. Quartz has a vitreous luster and conchoidal fracture.

Table 3.1
Recognizing Minerals in Igneous Rocks

Mineral	Properties
Potassium feldspar p. 17	Usually white or pink Two cleavages at 90° Equidimensional crystals
Plagioclase feldspar p. 17	Usually white (Na-plagioclase) or gray (Ca-plagioclase) Two cleavages at 90° Elongate crystals Striations
Quartz p. 18	Colorless to gray Glassy with conchoidal fracture Irregular crystals in intrusive rocks Equidimensional phenocrysts in extrusive rocks
Biotite p. 20	Shiny and black One perfect cleavage Thin crystals
Muscovite p. 20	Shiny and silvery white One perfect cleavage Thin crystals
Hornblende (amphibole) p. 19	Black with shiny, splintery appearance Two cleavages at 56° and 124° Elongate crystals
Augite (pyroxene) p. 19	Black, greenish black, or brownish black Vitreous, but rather dull, luster Two cleavages at 90° Blocky crystals
Olivine p. 20	Light green to yellow-green Glassy luster Small (few millimeters), equidimensional crystals

5. A pink feldspar is usually potassium feldspar; white or gray feldspars may be either potassium feldspar or plagioclase—if striations are present, it is plagioclase.

6. Cleavage and general appearance help to identify amphiboles and pyroxenes. Amphibole cleavages intersect at about 60° and 120°, whereas pyroxene cleavages intersect at 90°; amphiboles typically are elongate and have a splintery appearance, whereas pyroxenes look blocky.

Classification and Identification of Igneous Rocks

Igneous rocks are classified by their texture and mineral content. The classification scheme in Table 3.2 lists common textures in the left-hand column and typical mineral composition in the top row. Although rock names occupy distinct spaces in the table, understand that the dividing lines between rock types are artificially imposed; both mineral abundance and composition change gradationally from one rock type to the next.

When you are identifying a rock, it makes no difference whether you determine texture or mineral content first. Since texture is recognized more easily, first determine the average grain size and follow the top flowchart (texture) in Figure 3.6. This will lead you to one of the texture rows in Table 3.2. Next identify the major minerals and follow the lower flowchart (composition) in Figure 3.6. This will take you to the proper 'pigeonhole' in Table 3.2 that incorporates both texture and composition.

Some common igneous rocks are shown in Figures 3.7 through 3.16. **Remember, the photographs are representative examples only, and your samples may differ *considerably* in color and general appearance.** The **texture** and **mineral content** should be similar, so concentrate on those, not on color.

Crystallization of Magma and Bowen's Reaction Series

At high temperatures, a magma is completely liquid, but as the temperature drops, crystals begin to form. They don't all form at once, however. At first, crystals of only one or two minerals begin to grow. As the temperature continues to drop, these early-formed crystals may grow to become nicely shaped, they may react with the magma and be partially or wholly dissolved, or new minerals may begin to grow around them, preventing further growth. Continued crystallization increases the proportion of crystals and decreases the amount of liquid. Crystals grow up against one another and form an interlocking, crystalline network. Other minerals may crystallize with further temperature decreases; these will grow in the liquid remaining between earlier-formed crystals.

Because minerals form in a sequence, it may be possible to figure out that sequence by carefully studying the textural relations in the rock. This kind of process was done by N. L. Bowen during the early part of the 20th century, but he carried it a giant step further by duplicating the crystallization process in the laboratory. He paid particular attention to reactions between minerals as indicated by textural relations. For example, he commonly found rounded crystals of olivine surrounded by pyroxene (Fig. 3.17) and concluded that olivine formed first, then at a lower temperature, reacted with the liquid (magma) to form pyroxene. He called this a *discontinuous reaction,* because it results in the formation of a completely different mineral (pyroxene). As the pyroxene grows around the olivine, it sometimes prevents further reaction from taking place, and a partially reacted-upon, rounded olivine is left surrounded by pyroxene.

Bowen also observed that plagioclase feldspars gradually changed their chemical composition as magma temperature decreased and plagioclase crystallized. He attributed this to a *continuous reaction* between the magma and the growing crystals of plagioclase: as the temperature of the magma drops and plagioclase crystals

Table 3.2
Classification of Igneous Rocks*

Mineral Composition / Texture	Felsic	Intermediate	Mafic	Ultramafic
	• > 10% quartz, • > 50% feldspar with K-feldspar > Na-plagioclase, • ± < 15% biotite, muscovite, hornblende	• 0–10% quartz, • > 50% Na/Ca-plagioclase, • 0–10% K-feldspar • < 50% hornblende (especially), biotite, augite	• 20–85% Ca-plagioclase, • 15–50% augite, • 0–35% olivine	• Olivine, and/or pyroxene
Pegmatitic	Granite pegmatite	Diorite pegmatite	Gabbro pegmatite	
Coarse grained (phaneritic)	Granite	Diorite	Gabbro	Peridotite (dunite, if mostly olivine)
Coarse grained and porphyritic	Porphyritic granite	Porphyritic diorite	Porphyritic gabbro	
Fine grained (aphanitic)	Rhyolite	Andesite	Basalt	
Fine grained and porphyritic	Porphyritic rhyolite	Porphyritic andesite	Porphyritic basalt	
Fine grained and vesicular		Vesicular andesite	Vesicular basalt	
Pyroclastic	Rhyolite tuff or breccia	Andesite tuff or breccia	Basalt tuff or breccia	

Texture	Glass Equivalents of Felsic and Intermediate Rocks		Glass Equivalents of Mafic Rocks	
Glassy	Obsidian			
Glassy with many small vesicles	Pumice		Scoria (may have some crystals)	

*For simplicity, names of less common rocks are omitted and their spaces are left blank.

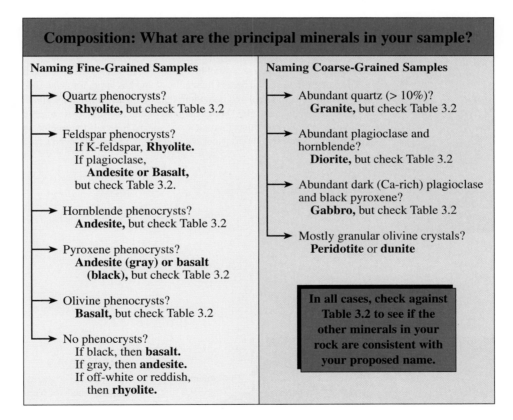

Texture: Start with the average crystal size.

Not visible or barely visible with a hand lens

→ Rock looks like glass?
   Glassy texture (obsidian)

→ Not glassy; grains are like fine or medium sandpaper?
   Fine-grained texture

→ Fine-grained with vesicles?
   Fine-grained, vesicular texture

→ Rock contains fragments?
   Pyroclastic texture (tuff)

→ Many small vesicles and black or red? **Scoria**

→ Light-weight, extremely vesicular, and light colored? **Pumice**

Visible (crystals larger than 1 mm) 1-mm squares

→ Most crystals larger than 3 cm?
   Pegmatitic texture

→ Most crystals smaller than 3 cm?
   Coarse-grained texture

→ Visible crystals (phenocrysts) in a fine-grained groundmass?
   Fine-grained, porphyritic texture

→ Distinctly larger crystals (phenocrysts) in a coarse-grained groundmass?
   Coarse-grained, porphyritic texture

Composition: What are the principal minerals in your sample?

Naming Fine-Grained Samples

→ Quartz phenocrysts?
   Rhyolite, but check Table 3.2

→ Feldspar phenocrysts?
   If K-feldspar, **Rhyolite.**
   If plagioclase,
   Andesite or Basalt,
   but check Table 3.2.

→ Hornblende phenocrysts?
   Andesite, but check Table 3.2

→ Pyroxene phenocrysts?
   Andesite (gray) or basalt (black), but check Table 3.2

→ Olivine phenocrysts?
   Basalt, but check Table 3.2

→ No phenocrysts?
   If black, then **basalt.**
   If gray, then **andesite.**
   If off-white or reddish,
   then **rhyolite.**

Naming Coarse-Grained Samples

→ Abundant quartz (> 10%)?
   Granite, but check Table 3.2

→ Abundant plagioclase and hornblende?
   Diorite, but check Table 3.2

→ Abundant dark (Ca-rich) plagioclase and black pyroxene?
   Gabbro, but check Table 3.2

→ Mostly granular olivine crystals?
   Peridotite or **dunite**

In all cases, check against Table 3.2 to see if the other minerals in your rock are consistent with your proposed name.

FIGURE 3.6
Flowcharts for the identification of igneous rocks. The top chart refers to igneous textures whereas the bottom one refers to composition. Both charts should be used together with the more comprehensive information in the text and Table 3.2.

Granite × 1

Photomicrograph
× 10

FIGURE 3.7

Granites are coarse-grained igneous rocks that make up the bulk of the large bodies of intrusive rock called batholiths. Most are light gray or pinkish, but gray and red granite are common too. Gray, glassy-looking quartz is an essential mineral, as is potassium feldspar. Potassium feldspar is commonly pink but may be white. If feldspars of two colors are present, the pink one is probably potassium feldspar, and the white or light gray one is probably sodium plagioclase. Most granites contain some black biotite or hornblende, and some contain muscovite. The fine-grained equivalent of granite is rhyolite. Some granites are very pretty when cut and polished, so they are quarried for use in monuments and buildings.

Granite pegmatite

FIGURE 3.8

Pegmatites are very coarse crystalline intrusive rocks that typically occur as small intrusions. The **granite pegmatite** shown here is light colored and has large, irregular-shaped crystals of gray quartz and crystals of salmon-colored potassium feldspar up to 30 cm long. Some varieties contain white, sodium-rich plagioclase. Big "books" of both biotite and muscovite mica are common. Some pegmatites contain rare minerals such as the silicate minerals tourmaline and beryl, both of which have several gem varieties (for example, emerald is a variety of beryl).

Diorite × 1

Photomicrograph
× 10

FIGURE 3.9

Diorite is a coarse-grained, intrusive rock usually made of white to light gray sodium-calcium plagioclase and black, splintery-looking hornblende, as shown here. Finely crystalline diorites look like a mixture of salt and pepper. Some diorites contain biotite as well as hornblende, and possibly a little quartz. The photomicrograph confirms the fully crystalline texture of this rock. Diorite is common in both large (batholith) and small intrusive bodies. Andesite is the fine-grained equivalent of diorite.

Gabbro × 1

Photomicrograph × 10

FIGURE 3.10

Gabbro is a coarse-grained intrusive rock composed of light to dark gray calcium plagioclase and black pyroxene. Olivine may be present as well. With so many dark minerals it can be tough to tell one from the other. However, in many gabbros, the plagioclase is easy to identify because the striations are readily visible on the large elongate crystals. Gabbro is the coarse-grained equivalent of basalt, and it commonly forms thin (dikes and sills) or irregular-shaped, small bodies of intrusive rock. The plagioclase in some gabbros causes a peculiar iridescent blue play of colors that makes the rock valuable for monument or building purposes.

Peridotite × 2

Photomicrograph × 10

FIGURE 3.11

Peridotite is generally a dark green to black rock, because its principal constituents are pyroxene and olivine. Peridotite consisting almost entirely of olivine is called *dunite,* which usually has the green color of olivine. In fact, the mineral sample of olivine you saw in the lab was probably a piece of dunite (see Figure 2.8). Peridotite rich in pyroxene may have fairly large crystals; in many peridotites, however, the crystals are small, though large enough to identify. Although peridotite is not abundant in the crust, it is the most abundant rock beneath the crust, in the upper part of the Earth's mantle.

Rhyolite × 1

Photomicrograph × 10

FIGURE 3.12

Rhyolite is a fine-grained extrusive rock that comes in a variety of colors. It is commonly light gray or pink, but red or even black rhyolite is not rare. Many rhyolites, like the one shown here, are porphyritic, so they are identifiable by the glassy quartz phenocrysts and white to salmon potassium feldspar phenocrysts in an aphanitic or even glassy groundmass. In the absence of phenocrysts, a light color suggests, but does not prove, rhyolite. The photomicrograph shows a few phenocrysts (gray) surrounded by a groundmass (black) with tiny crystals (brown to white). Granite is its coarse-grained equivalent.

Andesite × 1

Photomicrograph × 10

FIGURE 3.13

Andesite is the common extrusive rock of volcanoes formed at convergent plate boundaries. It is typically gray and porphyritic, with phenocrysts of white to light gray plagioclase and black hornblende or biotite. The photomicrograph shows a few phenocrysts (white, yellow, brown) surrounded by a glassy groundmass containing very tiny crystals. It is the fine-grained equivalent of diorite.

Basalt × 0.5

Rhyolite tuff × 1

Photomicrograph
× 10

FIGURE 3.14

Basalt is the most abundant extrusive igneous rock, forming the base of the seafloor and large plateaus on land. It is dark gray and fine grained, as in Figure 3.2, and can be vesicular, as shown here. Porphyritic basalt may have phenocrysts of calcium plagioclase, olivine, or both. Olivine phenocrysts are usually small (1 to 5 mm [0.04 to 0.2 in]) and are recognizable by their green color and glassy luster. Basalt is the fine-grained equivalent of gabbro.

FIGURE 3.15

Tuff is a textural name indicating a pyroclastic texture and particle size less than 6.4 cm (2.5 in). A similar rock with particles bigger than 6.4 cm (2.5 in) is a volcanic breccia. Compositional varieties of tuff or breccia are indicated by using rhyolite, andesite, or basalt as prefixes when possible; for example, rhyolite tuff, basalt breccia. Tuffs vary widely in appearance. Some are made entirely of compacted volcanic glass shards, others have abundant mineral crystals, and others consist of many angular fragments of extrusive rock, especially pumice; most commonly, all three components are present. Just as components and composition are variable, so is color, although many tuffs are light colored, like the rhyolite tuff shown here.

A. Pumice × 1

B. Scoria × 1

FIGURE 3.16

A. **Pumice** is the rock that floats because it has so many tiny holes in it. It is a froth of volcanic glass. Look at it with a hand lens to appreciate its character. Pumice is commonly, but not always, white or light gray. It is a good abrasive and is used in Lava® soap and in blocks used to scrape calluses off your feet or to clean griddles. B. **Scoria** is a basaltic rock with many small holes. You can think of it as being extremely vesicular with small vesicles. It is dark in color, ranging from black to reddish brown. It rarely looks vitreous, although microscopic examination of fresh samples shows that they formed as a bubbly glass. Although it looks as if it might float in water, it is usually too dense. Scoria is sold under the general name "lava rock" for outdoor decorative purposes.

grow, they continuously change their chemical composition and become richer in sodium.

Bowen summarized his studies with a diagram, now called *Bowen's Reaction Series,* which includes all the major rock-forming minerals of common igneous rocks (Fig. 3.18). The *discontinuous reaction series* contains the common iron- and magnesium-bearing silicate minerals. Plagioclase makes up the *continuous reaction series.* The three minerals at the bottom do not form by reaction. They simply crystallize last, sometimes in the sequence in which they are listed and sometimes simultaneously.

The reaction series shows that when a natural magma cools, certain minerals crystallize earlier and at higher temperatures than others. The common rock-forming minerals form in a regular sequence, with olivine and calcium-rich plagioclase at the top of the list, crystallizing at temperatures some 500°–600°C higher than quartz, which is at the bottom of the list. The crystalliza-

FIGURE 3.17

This photomicrograph shows rounded olivine crystals (green and red) surrounded by pyroxene (yellow). The early-formed olivine reacted with the magma to form some of the pyroxene that now surrounds it (× 10).

	Bowen's reaction series		Corresponding rocks		
			General	Intrusive	Extrusive
High temperature, first to crystallize	Olivine / Pyroxene / Amphibole / Biotite — Discontinuous series — Plagioclase feldspar Ca-rich / Continuous series / Na-rich		Mafic		
			Intermediate		
Low temperature, last to crystallize	Potassium feldspar ± Muscovite + Quartz		Felsic		

FIGURE 3.18

Bowen's Reaction Series. Minerals at top crystallize first from a magma at high temperatures (1200–1300°C); those at bottom crystallize last at lower temperatures (600–900°C). Iron- and magnesium-bearing silicate minerals are related by discontinuous reactions with the magma as temperature decreases, whereas plagioclase reacts continuously to form crystals progressively richer in sodium. For names of *Corresponding Rocks,* see Problem 5a.

tion temperatures of the plagioclase feldspars fall in the same range as the iron- and magnesium-bearing silicate minerals. Note, however, that the more calcium-rich plagioclases crystallize at higher temperatures than the less calcium- and more sodium-rich plagioclases. The most sodium-rich one crystallizes at the lowest temperature.

Cautionary note: Although the reaction series is instructive, composition and many other factors influence the precise crystallization history of a magma. The *Reaction Series* experiments focused on what happens when one type of basaltic magma cools. While the series summarizes the common order of crystallization and common reactions, the overlapping arrows in the discontinuous series suggest and allow for significant variations in the crystallization sequence.

Where and How Magmas Form

Most volcanoes are associated with tectonic plate boundaries (Fig. 3.19). What happens at plate boundaries that allows magmas to form? The first part of the answer involves the concept of **partial melting.** For example, what would happen if a rock made of quartz, potassium feldspar, sodium-plagioclase, biotite, and hornblende began to melt? We can use Bowen's Reaction Series to predict the results. The minerals lowest in the series, such as quartz and potassium feldspar, would melt at lower temperatures than minerals higher in the series. Imagine that the temperature got just high enough to melt the quartz and feldspar, but not high enough to melt the rest of the minerals. Because quartz and potassium feldspar are the main minerals in granite, the *partial melt* would have a felsic composition similar to that of granite. The liquid, because it is less dense than the surrounding rock, would migrate upward. Drops of liquid would eventually coalesce to form a body of magma, which would continue to move upward until it either erupted on the surface or crystallized below the surface.

The second part of the answer is that partial melting takes place at plate boundaries because it is here that rock either moves up toward the surface or down below the surface. For example, at *divergent plate boundaries* (where plates move away from one another), hot, buoyant peridotite (the main type of rock in the mantle; see Table 3.2) rises in the mantle as the plates move apart (Fig. 3.19). As it rises to shallower depths, pressure decreases until at some point it is low enough to cause the hot peridotite to partially melt. The partial melting of ultramafic peridotite produces a mafic magma. This magma rises and intrudes into the surrounding lithosphere to form gabbro or is extruded onto the seafloor as basalt (Fig. 3.19).

At *convergent plate boundaries* (where plates move toward one another), the cold oceanic seafloor is dragged down into the mantle under the overriding plate (Fig. 3.19). At some depth, the downgoing oceanic plate releases its water-rich fluids into the ultramafic rocks of the overlying mantle. The addition of water to hot rocks lowers the melting points of some minerals and partial melting takes

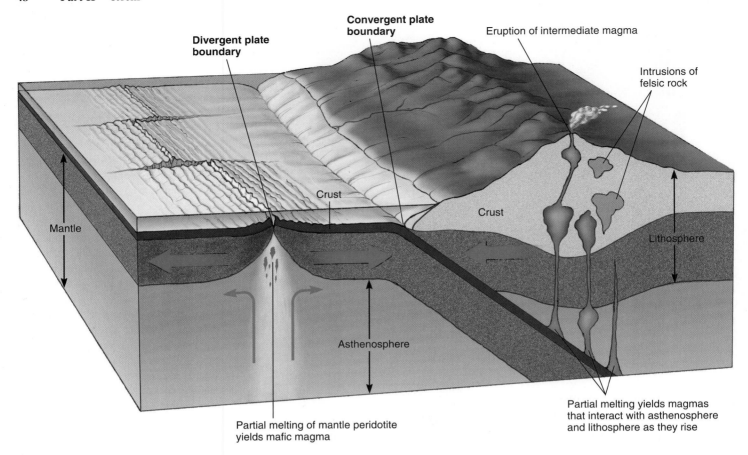

Divergent plate
boundary

Convergent plate
boundary

Eruption of intermediate magma

Intrusions of
felsic rock

Crust

Crust

Mantle

Lithosphere

Asthenosphere

Partial melting of mantle peridotite
yields mafic magma

Partial melting yields magmas
that interact with asthenosphere
and lithosphere as they rise

FIGURE 3.19

Most igneous activity takes place at divergent and convergent plate boundaries. Partial melting of peridotite mantle at divergent boundaries yields mafic magmas. Convergent boundaries are more complex; magmas formed by partial melting may interact with a variety of rocks to produce abundant intermediate and felsic magmas.

place. Mafic magmas are produced that rise through the mantle and enter the crust above. As these magmas rise, they may give off enough heat to partially melt the overlying rocks to produce a secondary magma, which also may rise in the crust. Because of the nature of partial melting, the new magma likely will be intermediate or felsic in composition.

Thus, all primary magmas start out as mafic partial melts of a peridotite mantle. If the magma can partially melt crustal rocks of different compositions, a wide range of magma compositions can be produced by secondary melting. This process, plus others that are beyond the scope of this lab manual, explain why we see so many different types of igneous rocks on Earth.

Your job in this lab is to name (classify) a wide range of igneous rocks. As you work through the lab, think about how these rocks actually formed. Keep in mind

that the intrusive rocks in your hands are the frozen remains of long-dead magma chambers, many of which probably fed ancient volcanic eruptions. As you handle the extrusive rocks, think about how some textures indicate relatively peaceful lava flows while others formed as the result of explosive eruptions that spelled death for anything living too close to the volcano. Understanding the origin of rocks is how you learn to "read" the rock record.

Hands-On Applications

When a geologist picks up a rock, she or he wants to know what kind of rock it is and how it formed. This information is the basis for most geological endeavors, whether it be a search for mineral resources or an environmental impact study. To "read" the rock, it is necessary to identify its constituents and interpret the basic textural relations. This lab will show you how to identify igneous rocks and how to use their textures to interpret their mode of formation. You also will learn how it is possible to make logical inferences or predictions from existing information. This is one of the most important things that geologists, or any scientists, do.

Objectives

If you are assigned all the problems, you should be able to:

1. Recognize the following textures in igneous rocks: glassy, aphanitic (fine-grained), phaneritic (coarse-grained), pegmatitic, porphyritic, vesicular, and pyroclastic.

2. Explain how each texture formed.

3. Distinguish between intrusive and extrusive igneous rocks on the basis of texture.

4. Recognize quartz, potassium feldspar, plagioclase, biotite, hornblende, augite, and olivine in an igneous rock.

5. Explain how igneous rocks are classified.

6. Identify igneous rocks.

7. Determine the relative temperature (high, intermediate, low) at which a common igneous rock formed, using Bowen's Reaction Series.

8. Relate the common igneous rocks to their plate-tectonic setting.

Problems

1. Which of the rock samples provided for you have the following textures? (Some of these may not be available to you.)

   Glassy:

   Fine grained (aphanitic):

   Coarse grained (phaneritic):

   Pegmatitic:

   Porphyritic:

   Vesicular:

   Pyroclastic:

2. Of the rock samples in Problem 1 with glassy, fine-grained, coarse-grained, pegmatitic, and porphyritic textures, which:

   a. Formed from the most rapidly cooled magma?

   b. Probably resulted from two stages of cooling—slow, then rapid?

   c. Crystallized from a water-rich magma?

3. Which of the rocks in the previous question are likely to be extrusive, and which are likely to be intrusive igneous rocks? Why?

4. Identify the samples of igneous rocks provided by your instructor. You may find it helpful to use the tear-out Igneous Rock Identification Worksheets, on the last pages of this chapter.

5. Use Bowen's Reaction Series (see Fig. 3.18) to answer the following.

   a. Based on the mineral content and texture of the following rocks, place them in the appropriate blank spaces under *Corresponding Rocks* in Figure 3.18: granite, gabbro, andesite, diorite, rhyolite, basalt. Which forms at the highest temperature, granite or gabbro? Andesite or basalt?

   b. Based on your knowledge of the colors of rock-forming minerals, are rocks at the top of Bowen's Reaction Series likely to be darker- or lighter-colored than those at the bottom? Which of the rocks in 5a are felsic? Intermediate? Mafic?

   c. If you identified quartz in a sample of igneous rock, what other minerals might you expect to find in the rock?

   d. If you identified olivine in a sample of igneous rock, what other minerals might you expect to find in the rock?

   e. Would you expect to find olivine and quartz in the same igneous rock? Explain your answer.

6. Most igneous rocks form at divergent and convergent plate boundaries. Of granite, gabbro, andesite, diorite, rhyolite, and basalt, which are most likely to form at a divergent boundary?

   At a convergent boundary?

7. Go to **Volcano World** (*http://volcano.und.edu/*, or link to it through *www.mhhe.com/jones6e*—see Preface). Use the links under *Volcanoes* to find information on Mount Rainier, a volcano near Seattle, Washington, and answer the following questions. Some links have **highlighted** terms in their text that lead to a glossary link. The glossary may also be useful for text that does not provide glossary links. Its address (at present) is *http://vulcan.wr.usgs.gov/Glossary/framework.html.*

   a. In what mountain range is Mount Rainier?

   b. Is Mount Rainier a shield volcano, a composite volcano (also called stratovolcano), or a cinder cone?

   c. Does it have a caldera at its summit?

   d. Volcanoes grow when the rate of addition of new volcanic materials exceeds the rate of erosion. Is Mount Rainier currently getting larger or smaller?

   e. When did Mount Rainier first become active?

   Click on *Radar Image of Mount Rainier* at the bottom of the page and answer the next two questions.

   f. When was the most recent eruption?

   g. Mount Rainier is considered a dangerous volcano. Why?

   Hit the *Back* button and click on the link to the *Cascade Volcano Observatory*. At the bottom of the page, under *Other Menus of Interest,* click on *Cascade Range Volcanoes and Volcanics Menu* (*http://vulcan.wr.usgs.gov/Volcanoes/Cascades/framework.html*). Click on *DESCRIPTION: Cascade Range Volcanoes and Volcanics* and answer the following questions.

   h. Which states host the Cascade Range volcanoes?

   i. List several other volcanoes in the Cascade Range.

   j. Explain, in terms of plate-tectonic theory, why Mount Rainier and the other Cascade volcanoes are there.

   k. What rock compositions (such as basalt or granite) are the most common?

   Another volcano website of interest is *http://satftp.soest.hawaii.edu/space/hawaii/virtual.field.trips.html,* which features virtual field trips of the Hawaiian Islands. If you would like to see very nice photomicrographs of igneous and metamorphic rocks, try *www.geolab.unc.edu/Petunia/IgMetAtlas/mainmenu.html.*

In Greater Depth

8. Indicate which chemical elements increase in abundance toward the bottom of Bowen's Reaction Series (refer to the chemical compositions given in Table 2.3).

9. Chemical analyses of representative extrusive rocks from the Cascade Range in Oregon are given in Table 3.3. Chemical analyses of rocks commonly are given in terms of oxides; that is, elements such as Si (silicon), Fe (iron), and Mg (magnesium) combined with oxygen (O). Notice that SiO_2 is by far the most abundant oxide and that it makes up from about 50 to 75% of these rocks. These values are typical of most igneous rocks. The amount of SiO_2 can be used to identify and name extrusive rocks, many of which are too fine grained to identify on the basis of mineral content. For one category of extrusive rocks, the following names are assigned on the basis of percent SiO_2:

Percentage SiO_2	Name of Extrusive Rock
<52	Basalt
52–57	Basaltic andesite
57–63	Andesite
63–70	Dacite
>70	Rhyolite

a. A common way to express a table of numbers such as Table 3.3 is to make a graph. A graph not only summarizes the data in a visually effective manner, it shows trends that otherwise may not be evident. In this case, you will make a graph to show how the oxides FeO, MgO, CaO, Na_2O, and K_2O vary with SiO_2. Figure 3.20 will be used for this purpose. Figure 3.20 already has the X-axis (horizontal) labeled as "Percent SiO_2" and the Y-axis (vertical) as "Percent Oxides". Examine the table with the weight-percent data and note that the SiO_2 numbers range from about 48 to 74%. To make an easily read graph, label the X-axis with values ranging from 45 to 80% along the whole bottom. The other oxide data show numbers ranging from 0.2 to 10.6%. Label the Y-axis with values ranging from 0 to 12%.

b. Now plot the data from the table. First plot SiO_2 (X-axis) versus FeO (Y-axis) for each sample. Connect the points and label the line at one end as "FeO". Then do the same with MgO, CaO, Na_2O, and K_2O.

c. Finally, to see how these data compare to the rocks you have seen in lab, draw vertical lines at each of the percent SiO_2 boundaries that divide each of the rock types: 52, 57, 63, and 70% SiO_2. In the spaces between these lines, label with the rock types listed above (basalt, basaltic andesite, etc.) along the top of the graph.

d. Which elements decrease with increasing abundance of SiO_2 and which ones increase?

Looking at Bowen's Reaction Series, what could explain the changes in the FeO, CaO, and MgO concentrations going from low to high SiO_2 lavas?

Why would the Na_2O and K_2O concentrations change?

e. Now that you can put a rock name to each sample, and using what you know from Bowen's Reaction Series, list the sample numbers in the order in which these rocks should have formed, from earliest to last.

Table 3.3
Chemical Analyses of Representative Extrusive Rocks from the Cascade Range, Oregon

Source: Data from D. J. Geist et. al., 1985, GPP (Geochemical Program Package), Center for Volcanology, University of Oregon.

	1	2	3	4	5	6	7	8
SiO_2	48.60	50.68	63.81	59.85	53.50	69.36	74.12	55.92
FeO	10.31	9.27	4.05	5.39	8.31	2.32	1.26	7.32
MgO	6.73	6.12	2.28	3.58	5.42	1.14	0.16	4.04
CaO	10.65	9.67	4.87	5.95	7.63	3.07	1.23	6.77
Na_2O	2.47	2.70	3.72	3.65	3.18	3.92	4.08	3.49
K_2O	0.25	0.48	1.96	1.28	0.73	3.02	4.47	0.97
Other	20.99	21.08	19.31	20.30	21.23	17.17	14.68	21.46

Numbers are weight percentages of constituents in the rock.

Percent Oxides

Percent SiO_2

FIGURE 3.20
Graph for Problem 9.

A. Shale × 1

B. Mudstone × 1

FIGURE 4.15

(A) **Shale** and (B) **mudstone.** The finest particles end up as components of these mud rocks. Shale is fissile; mudstone is not. Both tend to break up easily in outcrops, but mudstone is more likely to break into larger pieces with a conchoidal fracture. If rich in organic matter, shales and mudstones are black; if not, they are gray, red, or green. Some geologists distinguish two types of mudstone—claystone and siltstone—based on particle size. Siltstone feels gritty when chewed; claystone does not. (So, if you like, take a chew.) Mudstone and shale are by far the most abundant kinds of sedimentary rocks. They accumulate in the quiet water of floodplains, lagoons, and lakes, and offshore in shallow and deep marine environments.

A. Micritic limestone × 1

B. Crystalline limestone × 1

C. Oolitic limestone × 1

D. Fossiliferous limestone × 1

E. Coquina × 1

FIGURE 4.16

Limestone is made of calcite, so its hardness of 3 and ability to bubble readily in acid are distinctive. Only a few of the many types of limestone are mentioned here. *Micritic* or *microcrystalline limestone* (A) has such tiny crystals that it often forms smooth fracture surfaces. Most samples represent the recrystallized remains of billions of microscopic particles released by the decay of calcareous algae. *Crystalline limestone* (B) is made of small but recognizable interlocking calcite crystals. Many samples formed when coarse bioclastic fossil debris was extensively recrystallized during or after lithification. Few obvious fossils remain. *Oolitic limestone* (C) (the *oo* is pronounced like the two o's in moo) consists of sand-sized spheres of calcite (called *ooids*) cemented together. The interior of an ooid shows concentric layers of calcite formed around a central tiny bioclastic particle. Ooids form like tiny snowballs in warm shallow waters in response to wave or tidal currents that wash them back and forth. *Fossiliferous limestones* (D) contain abundant calcareous (made of calcium carbonate) fossils or fossil fragments that are cemented together with microcrystalline or crystalline calcite. Such limestones frequently formed in shallow continental shelves; many contain impurities in the form of clay and silt-sized clastic particles. *Coquina* (E) consists almost entirely of poorly cemented, coarse, bioclastic debris. There is little or no fine-grained matrix so the rock is very open and porous. Coquinas generally form near beaches where waves wash away the finer grains. *Chalk* is a white, soft limestone composed of microscopic calcareous fossils that accumulate in deeper ocean waters. In general, most limestones form in warm, shallow seas far from shore, where the water is clear and free from land-derived detritus. Although today confined to small parts of the outer continental shelves, the geologic record of limestones tells us that in the past such seas covered vast areas of the continents.

Dolostone × 0.5

FIGURE 4.17

Dolostone is made of the mineral dolomite. It is distinguishable from limestone by its lesser reaction with acid. It is sometimes necessary to use a hand lens to see that it actually bubbles, or to scratch the rock (to make a powder) before testing with acid. Dolostone has a crystalline texture and commonly is light gray or buff. A common, but not invariable, feature is the presence of crystal-lined cavities, or *vugs*. The sample shown here contains a crystal of clear quartz (a "Herkimer diamond"). Most dolostone is formed by replacement of limestone.

Chert × 1

FIGURE 4.18

Chert (shown here), **flint,** and **jasper** are made of microcrystalline quartz. They break with a conchoidal fracture and have smooth surfaces. They most resemble micritic limestone, but their hardness of 7 gives them away. See Table 4.3 and Figure 2.4C, D, and E for characteristic colors. Some chert forms by replacement of limestone and may retain some of the textural elements, such as fossils or ooids. Most chert forms by biochemical extraction of silica from ocean water. Such chert consists of microscopic plant (diatom) or animal (radiolaria and sponge spicules) fossils.

Rock gypsum × 1

FIGURE 4.19

Rock gypsum is usually white, light gray, or pale shades of red or orange. It has a crystalline texture and can be scratched with your fingernail. Dark impurities may be present as layers or outlines around irregular areas. Gypsum is an evaporite mineral; when seawater evaporates, gypsum is one of the minerals that precipitates. Thick deposits accumulate in bays in arid regions that have restricted access to the open ocean; deposition also occurs in salty inland seas or lakes.

Rock salt × 1

FIGURE 4.20

Rock salt is made of halite. It has a crystalline texture and is white, light gray, or tinged with red or orange. It forms in the same manner and depositional environment as does gypsum. The reddish mineral here is sylvite (KCl).

Coal × 1

FIGURE 4.21

Coal is brown to black, depending on the amount of carbon it contains. Brown to brownish black *lignite* contains less carbon than black *bituminous coal,* which, in turn, contains less carbon than metamorphic coal, *anthracite*. Coal is noncrystalline and has a luster that ranges from dull to vitreous as carbon content increases. Coal forms by burial and compaction of vegetation that accumulated in swamps.

Hands-On Applications

This lab continues the process of learning to identify and interpret rocks by focusing on sedimentary rocks. You also will learn how observation, measurement, and comparison help to establish general relations between characteristics of rocks and their history of transport and depositional environment.

Objectives

If you are assigned all the problems, you should be able to:

1. Determine whether a sedimentary rock has a clastic, bioclastic, or crystalline texture.

2. Distinguish grains from matrix in a clastic sedimentary rock.

3. Identify a sedimentary rock and the major minerals contained in it.

4. Estimate grain size, degree of rounding of grains, and degree of grain sorting of a clastic sediment or sedimentary rock.

5. Recognize the following sedimentary structures in a rock sample or a picture: stratification, cross-bedding, ripple marks, graded bedding, and mud cracks.

6. Study selected sedimentary rock samples and suggest possible depositional environments in which they may have formed.

Problems

1. To help you learn to recognize the principal textures of sedimentary rocks, sort the samples provided for you into the following textural categories:

   Clastic

   Bioclastic

   Crystalline

   Fine-grained clastic or microcrystalline

2. Identify the samples of sedimentary rocks provided by your instructor. You may find it helpful to use the tear-out Sedimentary Rock Identification Worksheets at the end of this chapter.

3. Examine the sample of loose sediment provided for you, using a hand lens or a binocular microscope.

   a. Identify the major minerals and estimate their proportions.

   b. By comparison with the roundness scale (Fig. 4.1B), determine the roundness of the grains. Are some minerals rounder or more angular than others? If so, which are most angular and which are most rounded? Suggest an explanation for the similarities or differences.

   c. Use a metric ruler or graph paper with 1 mm squares to measure the diameters of several grains in millimeters. What is the range in size, and what sizes are the most prevalent? What grain-size names (for example, sand or granules) best describe the sediment (see Fig. 4.1A)?

d. Describe the degree of sorting. For the purposes of this lab, consider the sediment to be well sorted if most grains have the same grain-size name, and poorly sorted if they do not.

e. Make the unlikely assumption that all of the grains were freed from the source area by mechanical weathering (that is, no chemical weathering took place). Based on the mineral content of the sediment, what igneous rock(s) could have been the source of this sediment?

f. What was the probable transporting agent (wind, water, ice) for the materials in the sediment? Explain.

g. Is the sediment detrital or chemical/biochemical? Why?

h. What kind of sedimentary rock would be formed by lithification of this sediment?

4. Examine samples of the following sedimentary structures, and answer the questions.

   a. Graded bedding.

   (1) Before looking at the rock sample, shake a jar of water containing a mixture of clay, silt, sand, and granules; let it sit on the lab bench while you do something else for 5 or 10 minutes.

   (2) Look at the sediment that has settled on the bottom of the jar. How are different-size grains distributed? Are similar-size grains segregated in layers, or are they distributed randomly throughout the sediment?

   If in layers, are the layers parallel or perpendicular to the bottom of the jar?

   Does the grain size change in any regular manner from top to bottom or from one side of the jar to the other?

   Is there still sediment in suspension that has not settled? If so, what grain-size term would describe it most aptly?

   (3) Next, look at the rock sample, paying particular attention to the distribution of its grains. Make a rough sketch of the sample showing stratification and emphasizing grain-size differences within or across strata. By analogy with the sediment in the jar, determine the original top of the sample and label it in your sketch.

   b. Ripple marks. Sketch the sample to show the shape of the ripple marks in cross section, and identify the type of ripple mark. For current ripple marks, indicate with an arrow the current direction at the time the ripples formed. For oscillation ripple marks, label the top and bottom if you can. Explain your answer.

c. Cross-stratification. Sketch the sample to show the strata in cross section and indicate with an arrow the current direction at the time of deposition. Explain your answer.

d. Mud cracks. Sketch the original top of the sample. How did you know which side was the original top?

5. Go to *www.kaibab.org/geology/gc_layer.htm* (or link to it through *www.mhhe.com/jones6e*—see Preface), which contains descriptions of the rocks exposed in the Grand Canyon.

   a. What kinds of Paleozoic sedimentary rocks occur in the Grand Canyon (types of rocks, not names of formations)?

   b. At the bottom of the web page is a drawing summarizing the rocks seen in the Grand Canyon. Use the maximum thickness given for each of the Paleozoic formations to calculate the relative proportions of the following rock types: carbonates (limestone plus dolostone), shale, and sandstone.

   c. A global average of the relative proportions of the different rock types reveals the following proportions: carbonate = 15%; shales = 65%; sandstones = 20%. When you gaze out across the Grand Canyon (see links below), are you seeing a set of rocks comparable in composition to the world average?

   d. Based on their descriptions and Table 4.2 in this lab manual, what are the depositional environments of the following:

     (1) Kaibab Limestone

     (2) Coconino Sandstone

     (3) Hermit Shale

(4)   Supai Formation

(5)   Temple Butte Limestone

You may be interested in seeing more of the Grand Canyon. Try *www.kaibab.org* and click on *armchair travelers*. You can take a virtual *13 Day River Trip,* make a *Virtual Visit,* and explore several other links showing photos of the many things you can see in and around the Grand Canyon.

In Greater Depth

6. Following are descriptions of some common sedimentary rocks, possibly similar to ones provided in lab. Using the descriptions, suggest possible depositional environments for each (see Fig. 4.2 and Table 4.2).

    a. Reddish rock made principally of angular to sub-rounded, coarse sand- to granule-size grains of quartz and feldspar. In the field, structures like those in Fig. 4.5 are evident, and the sandstone is seen to be interbedded with pebble and cobble conglomerate.

    b. Gray rock with numerous fossil shells cemented firmly together (similar to Fig. 4.16D). Rock readily fizzes with dilute HCl.

    c. White rock with crystalline texture. Very soft, does not fizz with acid or taste distinctive.

    d. A conglomerate with grain sizes ranging from clay to boulder. No apparent bedding, unsorted; particles range from angular to rounded, and some have flat, scratched faces.

    e. Light tan rock with well-rounded, 0.5 mm (0.02 in), frosted quartz grains. Well sorted, with large crossbeds.

    f. Black, shiny rock with some poorly preserved plant fossils.

    g. Reddish-gray, very fine-grained, detrital rock, with features like those in Fig. 4.7.

    h. Dark gray rock that looks somewhat like basalt, but with a hand lens, the texture appears to be clastic, not crystalline. Some white feldspar grains are present, as are dark grains of something unidentifiable; the matrix is too fine to recognize anything. One sample has thin beds that show a progressive decrease in grain size going from bottom to top.

7. Sedimentary geology is most interesting when you examine a whole series of rocks deposited in layers one on top of the other because, like pages in a book, they tell you a story. On a hike up a long hill you find the following succession of horizontal rock layers: (1) At the bottom are gray shales containing plant fossils and a thin layer of coal. (2) Then comes a transition to a quartz sandstone displaying abundant crossbeds and even some ripple marks. The grains are medium-sized at the base but become fine near the top. (3) This is followed by a transition to more shales, but no plant fossils are visible and there is no coal. Instead, there are ancient seashells of various types. (4) Finally, at the top of the hill come various layers of gray and light tan limestones, some of which display various marine fossils, including abundant corals. How did the environments of deposition change at this locality? What one thing could have occurred to cause these changes?

Sedimentary Rock Identification Worksheet

Sample Number	Texture Clastic = C Crystalline = X Fine-grained = F	Mineral Composition	Sorting and Roundedness (if applicable)	Fossils (leaves, clams, corals, etc.)	Condition of Fossils (largely whole or mostly broken up?)	Rock Name	Inferred Depositional Conditions*

*Grain size indicates current strength. Gravels = very strong currents or near sediment source; sands = strong currents; shales/mudstones = quiet conditions. Fossils may indicate a land or marine setting. Limestones and dolostones: most are marine and form in middle to outer shelf (including reefs); grain size indicates whether they experienced strong, moderate, or gentle currents from tides or waves. Broken fossils indicate higher currents than whole fossils. See Table 4.2 and text for more details on these and other depositional settings.

Sedimentary Rock Identification Worksheet

Sample Number	Texture Clastic = C Crystalline = X Fine-grained = F	Mineral Composition	Sorting and Roundedness (if applicable)	Fossils (leaves, clams, corals, etc.)	Condition of Fossils (largely whole or mostly broken up?)	Rock Name	Inferred Depositional Conditions*

*Grain size indicates current strength. Gravels = very strong currents or near sediment source; sands = strong currents; shales/mudstones = quiet conditions. Fossils may indicate a land or marine setting. Limestones and dolostones: most are marine and form in middle to outer shelf (including reefs); grain size indicates whether they experienced strong, moderate, or gentle currents from tides or waves. Broken fossils indicate higher currents than whole fossils. See Table 4.2 and text for more details on these and other depositional settings.

CHAPTER 5

Metamorphic Rocks

Materials Needed

- Pencil
- Hand lens
- Dilute hydrochloric acid
- Glass or knife to test hardness
- Samples of metamorphic rocks to be identified

Introduction

As a result of heat and/or pressure, rocks can undergo a process called metamorphism. Metamorphism can yield new crystals of the same mineral or, by reorganization of chemical constituents, crystals of new minerals. In the process new textures are developed. The characteristics of the original rock, and the type, intensity, and duration of metamorphism, determine how drastic the change is. This chapter will introduce some of the common varieties of metamorphic rocks, illustrate their typical textural and mineralogic features, show how they relate to the original or parent rock, and introduce the idea of metamorphic zones.

In some ways, metamorphic rocks are the most mysterious of the three major types of rocks. We can observe volcanoes and see how some igneous rocks form, and watch as sediment accumulates on the Earth's surface. But metamorphic rocks typically form well below the surface where we can't see what happens. To understand them, we must infer what goes on, or run experiments in the lab at high temperatures and pressures. We now think we understand how and under what conditions they form and can, in turn, use actual metamorphic rocks to understand what happens deep beneath the Earth's surface, especially in the cores of mountain belts and underneath volcanic regions, and to unravel the geologic past represented by metamorphic rocks found around our planet.

Metamorphic rocks are crystalline rocks that form from other rocks. **Metamorphism** takes place at varying depths within the Earth's crust, where both temperature and pressure are higher than at the surface. The results of metamorphism include the creation of new minerals, the development of bands or layers of like minerals (for example, layers of quartz or mica), and the parallel alignment of new and old mineral crystals.

Conditions of Metamorphism

Metamorphism takes place at temperatures and pressures that fall between those in which sediments are lithified and those in which rocks begin to melt to form magmas. Figure 5.1 shows approximate lower and upper boundaries of metamorphism. A rock will be metamorphosed if temperature and pressure conditions place it within the areas labeled "metamorphism" in Figure 5.1, and if there is sufficient time for the appropriate reactions to occur. Some rocks, such as granite or quartz sandstone, are not affected much by metamorphism until temperatures and pressures are well within the field of metamorphism on Figure 5.1; others, such as porous tuffs consisting of shards of volcanic glass, are highly reactive and undergo changes at temperatures and pressures near those of the field labeled *lithification of sediment*. Similarly, melting begins at dif-

ferent temperature-pressure conditions for different rocks.

Water and such fluids as carbon dioxide are generally present during metamorphism, because they are either contained in the rocks undergoing metamorphism or are released by metamorphic reactions. These fluids are important in the metamorphic process. Among other things, they allow ions to move about more readily, thereby speeding up the chemical reactions that enable the recrystallization and growth of mineral crystals. Without them reactions may not take place.

Types of Metamorphism

There are two common types of metamorphism: regional and contact.

Regional metamorphism occurs over areas of hundreds or thousands of square kilometers. Regionally metamorphosed rocks were buried beneath thick accumulations of other rocks at some time during their history (Fig. 5.2). During this period of burial, they were subjected to the higher temperatures and pressures found beneath the surface (Fig. 5.1). The weight of the overlying rocks provided most of the pressure, but collisions between tectonic plates, such as occurs at convergent boundaries, also creates pressure directed from the side. It is

Texture: Is the rock foliated or nonfoliated?

Foliated

- Slaty foliation (with crystals too small to see): **Slate**

- Phyllitic foliation (with barely visible crystals): **Phyllite**

- Schistose foliation (with crystals large enough to identify): **Schist**

- Gneissic foliation (layers of visible crystals mm to cm thick): **Gneiss**

- Enlarged pods of quartz/feldspar in the midst of swirling light and dark bands: **Migmatite**

Composition: What minerals are present in your foliated rock?

Names of foliated rocks are modified when possible by mineral names, as in quartz-biotite schist. Some use igneous rock names to describe gneisses with similar mineral compositions (e.g., granititic gneiss).

Nonfoliated or weakly foliated.

Composition: What is the non-foliated/ weakly foliated rock made of?

- Chiefly calcite or dolomite (H = 3-4, fizzes in acid): **Marble**

- Chiefly quartz (scratches glass): **Quartzite**

- Dark gray to black, shiny, relatively light weight, conchoidal fractures: **Anthracite coal**

- Gray, dark gray, greenish, or black: **Greenstone, amphibolite, soapstone, serpentiinite,** or **hornfels. See Table 5.2.**

- Somewhat recrystallized versions of conglomerate, graywacke, or chert: **Metaconglomerate, metagraywacke,** or **metachert. See Table 5.2.**

FIGURE 5.13 (Above)
Flowchart for the identification of metamorphic rocks. Foliated rocks are identified following the charts on the left while nonfoliated rocks follow the charts on the right. These charts should be used together with the more comprehensive information in the text and Table 5.2.

they also may contain non-igneous minerals such as clay or calcite. Sedimentary rocks with crystalline textures are composed of minerals, such as calcite, dolomite, gypsum, or halite, not usually found in igneous rocks. Foliated metamorphic rocks are distinguished from igneous and sedimentary rocks by foliation. Some contain typically metamorphic minerals like garnet, staurolite, or kyanite. Nonfoliated metamorphic rocks usually consist mostly of one mineral, such as quartz or calcite.

Some specific "problem rocks" are the following:

> **Shale vs. slate.** Because a slate forms by low-grade metamorphism of a shale, these two rocks are very similar. In general, a slate is shinier and a bit harder and tougher than a fissile shale. Thin pieces of slate tend to "ring" when dropped on a table, whereas shale tends to "thunk."

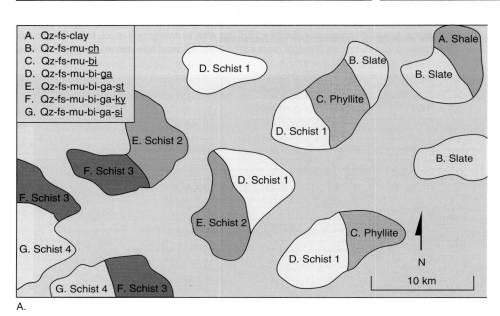

A.
- A. Qz-fs-clay
- B. Qz-fs-mu-<u>ch</u>
- C. Qz-fs-mu-<u>bi</u>
- D. Qz-fs-mu-bi-<u>ga</u>
- E. Qz-fs-mu-bi-ga-<u>st</u>
- F. Qz-fs-mu-bi-ga-<u>ky</u>
- G. Qz-fs-mu-bi-ga-<u>si</u>

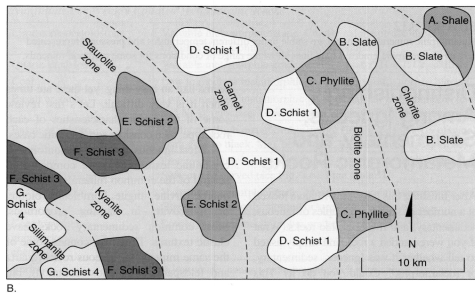

B.

FIGURE 5.14 (Left)
A. Map of outcrops indicating rock types and mineral assemblages. Qz = quartz; fs = feldspar; mu = muscovite; ch = chlorite; bi = biotite; ga = garnet; st = staurolite; ky = kyanite; si = sillimanite. Index minerals are underlined. B. Same map showing isograds and mineral zones.

Quartz sandstone vs. quartzite. Like shale and slate, a complete range, or gradation, can be seen from sandstone to quartzite, depending on the degree of metamorphism. One way to distinguish the two is based on the way the rock breaks. If the break cuts through individual quartz grains, the rock generally is quartzite; if it breaks around grains, it is quartz sandstone.

Limestone or dolostone vs. marble. Again, a complete range exists from limestone/dolostone to marble, depending on the degree of metamorphism. Marbles have readily distinguishable crystalline textures, with individual crystals large enough to see rather easily. Although some limestones have crystalline textures, the crystal size generally is quite small. Sometimes, however, it is nearly impossible to distinguish limestones from marbles in hand specimens. They usually can be distinguished in the field by examining the surrounding rocks; if they are sedimentary, it is likely a limestone; if they are metamorphic, it is likely a marble. Marbles are *not* harder than limestones; they are equally hard. Why?

Graywacke vs. basalt. A dark gray, fine-grained graywacke may be difficult to distinguish from basalt. However, with a hand lens, you generally can see the clastic texture in graywacke and the crystalline texture in basalt. Graywackes may contain quartz, whereas basalts do not.

Mudstone vs. tuff. Some tuffs are so fine grained that they look like mudstones. Tuffs are commonly lighter in color and weight than mudstones, but examination with a microscope may be necessary to distinguish the two.

Conglomerate or sedimentary breccias vs. volcanic breccia. Volcanic breccias or tuffs with pyroclastic fragments may look like conglomerate or sedimentary breccia. The volcanic rocks usually contain only fragments of other volcanic rocks, whereas the particles in conglomerate and sedimentary breccia may be any rock type or a mixture of rock types. Tuffs commonly contain small crystals of common igneous minerals and almost always contain tiny bits of broken volcanic glass or shards (these may be difficult to see). Many tuffs and some volcanic breccias are lighter weight than conglomerates.

Metamorphic Reactions

Most metamorphic reactions are between two or more solid minerals and occur as the result of an increase in temperature, pressure, or both. They result in the formation of new minerals and, in most cases, the release of fluids such as water or carbon dioxide. When the fluids are gone, the reaction cannot be reversed and the minerals formed at the highest temperature and pressure are preserved. Metamorphic reactions that have been studied in the laboratory can be used to estimate the actual conditions under which metamorphic rocks formed.

An example of a metamorphic reaction is illustrated in Figure 5.15. That figure shows the temperature and pressure (or depth) conditions at which the reaction *muscovite + quartz = potassium feldspar + sillimanite + H₂O* can occur. Muscovite and quartz can occur together without reacting, as long as the combination of temperature and pressure is to the left of the reaction curve; for example, at 500° and 3 kilobars. If the temperature at that pressure were higher, so that it was to the right of the reaction curve (for example, 700°), then muscovite and quartz could not occur together without reacting to form potassium feldspar, sillimanite, and water. When the rock cools again, the reverse reaction (forming muscovite + quartz) generally does not take place in nature, because the water necessary for the reaction has escaped. Because water and other fluids are driven off during metamorphism, the minerals present in metamorphic rocks generally are those formed at the highest temperature and pressure conditions that were attained.

It is sometimes possible to find rocks that contain both products and reactants. For the reaction shown in Figure 5.15, that rock would contain quartz, muscovite, potassium feldspar, and sillimanite. Such a rock would have been subjected to conditions right on the reaction curve (for example, about 630° at a pressure of 3 kilobars). If a series of outcrops with the same products and reactants could be located, they would all represent the same metamorphic grade. A line on a map connecting them would be an **isograd,** (see Fig. 5.14B), a line of equal metamorphic grade, generally defined by a specific reaction or the appearance of a specific mineral.

By using numerous reactions like the one shown in Figure 5.15, it is possible to determine the temperature and pressure conditions that must have existed at the time of metamorphism. Because pressure, especially, can be related approximately to depth (see Fig. 5.1), it is also possible to figure out how deep the rocks must have been when they were metamorphosed.

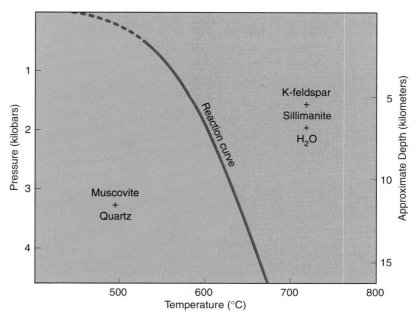

FIGURE 5.15

The reaction *muscovite + quartz = K-feldspar + sillimanite + H₂O* occurs when the combination of temperature and pressure (depth) fall on the reaction curve. If temperature and pressure are left of the curve, muscovite and quartz are present; if they are right of the curve, K-feldspar, sillimanite and H₂O are present.

Hands-On Applications

This lab continues the application of the scientific method to the study of rocks. Problems focus on identification and interpretation based on observation of metamorphic rock samples. You will see how correct identification will enable you to infer the parent of the rock you have identified and the approximate temperature-depth conditions during metamorphism.

Objectives

If you are assigned all the problems, you should be able to:

1. Identify the major minerals in metamorphic rocks.

2. Identify a metamorphic rock.

3. Distinguish between high- and low-grade metamorphic rocks.

4. Distinguish between foliated and nonfoliated metamorphic rocks.

5. Distinguish among slaty, schistose, and gneissic foliations.

6. Suggest a probable parent for a given metamorphic rock.

7. Determine whether a rock is igneous, sedimentary, or metamorphic.

8. Construct isograds on a map, given the appropriate data.

Problems

1. Metamorphic rocks are subdivided into two major textural groups: FOLIATED and NONFOLIATED. Divide the samples provided for you into these two categories.

   Foliated

   Nonfoliated

2. Which of the samples with foliated textures have the following specific types of foliation?

   Slaty

   Phyllitic

   Schistose

   Gneissic

3. Many foliated metamorphic rocks form by the metamorphism of shale; the more intense (or *higher grade*) the metamorphism, the coarser the crystals contained in the rocks.

   a. Select from among your samples those that may have been derived from shale, and arrange them according to metamorphic grade (see Table 5.2).

   Lowest grade:

Intermediate grade:

Highest grade:

b.   Assuming that the rocks you selected formed in the vicinity of a magmatic zone at a convergent plate boundary, what were the approximate temperatures and depths of each grade at the time of metamorphism (see Fig. 5.1)?

4. What are the *parent rocks* of the samples specified by your instructor (see Table 5.2)?

5. Identify the samples of metamorphic rocks provided for you. You may find it helpful to use the tear-out Metamorphic Rock Identification Worksheet, which is at the end of this chapter.

6. How can you distinguish between the following similar-looking rocks?
    a.   Marble and quartzite

    b.   Slate and phyllite

    c.   Schist and gneiss

In Greater Depth

7. When confronted with scattered outcrops of a variety of metamorphic rocks, a geologist must try to make some sense of their distribution in order to understand their origin. One of the first things to do is to make a map of the outcrops and indicate what rocks and minerals are present in each.

    Figure 5.16 shows large outcrops of several kinds of rocks, metamorphosed to varying degrees. Between the outcrops the rocks are covered with soil and other surficial sediment. The parent rocks for these metamorphic rocks are identified by various colors. The numbers on the outcrop refer to the samples described below the map. Where applicable, mineral names modify rock names and indicate the metamorphic zone (e.g., see Fig. 5.14B).

    a. Write abbreviated sample descriptions next to the sample numbers on the map. For example, next to 13 write "bt sch" for biotite schist.

    b. Where on the map is the metamorphism most intense? Least intense?

    c. Which parent rocks best record differences in metamorphic grade?

    Which parent rocks are least helpful for determining metamorphic grade?

    d. Using the index minerals as a guide, draw isograds (lines of equal metamorphic guide or intensity) on the map and label the mineral zones between them.

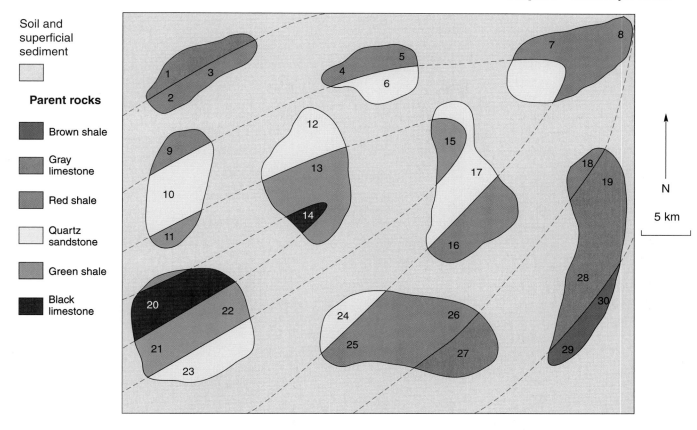

Soil and superficial sediment

Parent rocks

■ (dark)	Brown shale
■ (gray)	Gray limestone
■ (medium)	Red shale
□ (light)	Quartz sandstone
■ (medium-gray)	Green shale
■ (black)	Black limestone

N

5 km

Sample no.	Description	Sample no.	Description
1	Limestone	16	Garnet schist
2	Chlorite phyllite	17	Quartzite
3	Biotite schist	18	Staurolite schist
4	Garnet schist	19	Marble
5	Staurolite schist	20	Limestone
6	Quartzite	21	Green shale
7	Kyanite schist	22	Chlorite slate
8	Kyanite schist	23	Quartz sandstone
9	Chlorite phyllite	24	Quartzite
10	Quartzite	25	Chlorite phyllite
11	Chlorite slate	26	Biotite schist
12	Quartzite	27	Marble
13	Biotite schist	28	Marble
14	Marble	29	Garnet schist
15	Garnet schist	30	Garnet schist

FIGURE 5.16

This map and rock descriptions are to be used with Problem 7. The map shows the distribution of outcrops with different colors representing different parent rocks. The numbers correspond to the samples described in the list below the map. The solid lines between different colors show where different rock bodies are *observed* to be in contact with each other. The dotted lines indicate where these contacts are *inferred* to be under the soil and surficial sediment.

Metamorphic Rock Identification Worksheet

Sample Number	Texture	Mineral Composition	Other Properties (e.g., color)	Rock Name	Metamorphic Grade/ Probable Parent Rock(s)

Metamorphic Rock Identification Worksheet

Sample Number	Texture	Mineral Composition	Other Properties (e.g., color)	Rock Name	Metamorphic Grade/ Probable Parent Rock(s)

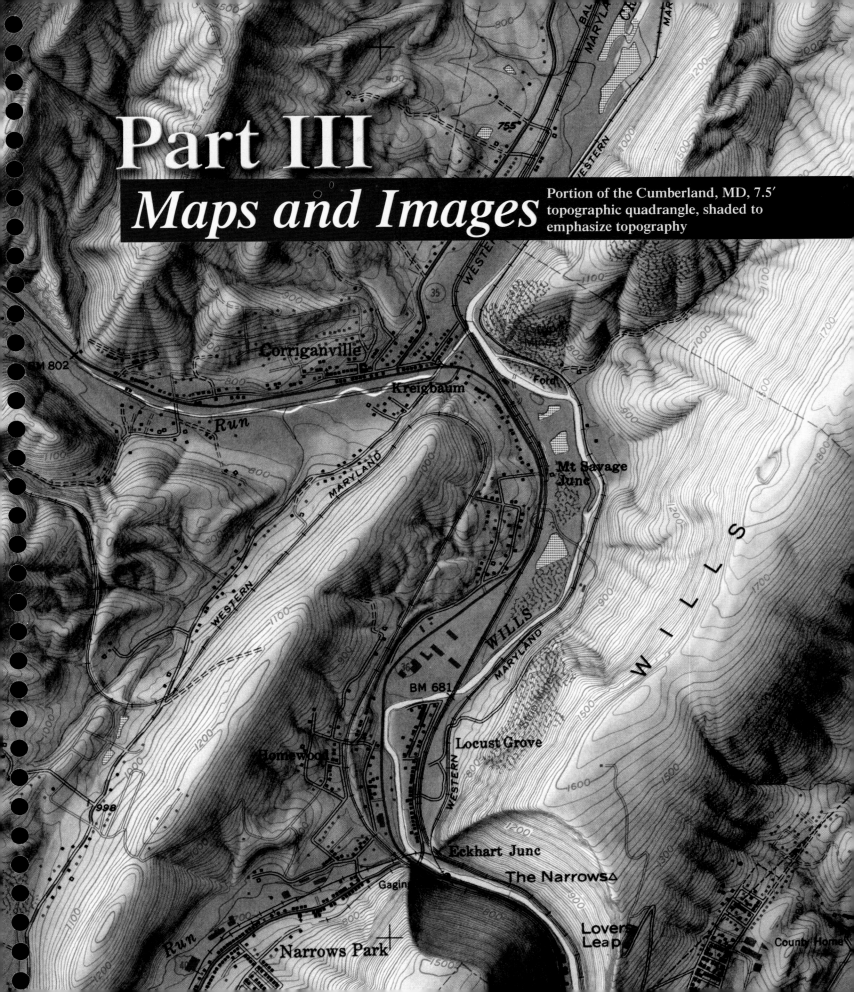

Part III
Maps and Images

Portion of the Cumberland, MD, 7.5′ topographic quadrangle, shaded to emphasize topography

Eckhart Junc
The Narrows△
Gaging Sta
Lovers
Leap
County Home
Narrows Park
Run

CHAPTER 6

Topographic Maps and Digital Elevation Models

Materials Needed

- Pencil and eraser
- Metric ruler
- Calculator

- Topographic quadrangle map (provided by your instructor)

Introduction

The topography of the Earth holds endless fascination for geologists and others who love the natural world. **Topography** refers to the hills, valleys, and other three-dimensional landforms on the Earth's surface. **Bathymetry** refers to similar features located beneath the sea.

Landscapes are interesting because they reflect the long-term action of erosional forces, such as streams, glaciers, and pounding waves at the beach, and differences in how easily the underlying rocks erode. By "reading" a landscape, geologists discover rock structures hidden beneath the soil, infer long sequences of past landscapes, and see that the land has uplifted or subsided. In the eyes of a geologist, a three-dimensional landscape gains the fourth dimension of time. Archaeologists familiar with natural landforms become adept at spotting unnatural features, which helps them discover sites of ancient human activity.

As you will learn in this chapter, landscapes are conveniently visualized with the aid of maps and digital elevation models. **Topographic maps** (Fig. 6.1A) precisely define the shape of landforms using topographic contours, which we'll introduce in the next section. **Digital elevation models (DEMs)** plot a high-resolution

A.

B.

FIGURE 6.1

A. Topographic map showing Meteor Crater, Arizona. Each brown contour line traces a specific elevation above sea level. B. Digital elevation model (DEM) of Meteor Crater, Arizona. Colors vary according to elevation (yellow high, green low). Scale: about 1 inch per half mile.

grid of elevations using different colors to indicate different elevations (Fig. 6.1B). DEMs are like shaded relief maps.

Topographic Maps

Topographic maps show the size, shape, and distribution of landscape features using contour lines, shading, coloring, or, especially on antique maps, short, closely spaced lines that schematically indicate mountains or steep slopes. We'll focus on contour lines because they most pre-

cisely depict the third dimension. Figure 6.2 compares the topography of an area with its contour map. You can also relate contours to topography in the shaded topographic map shown in the section opener (p. 93 and in Figure 6.1).

Contour Lines

A **contour line** is a line on which all points have the same elevation. It is shown in brown in Figure 6.1A. The elevation or altitude of a point on Earth is the vertical distance between that point

FIGURE 6.2

The area sketched in the top diagram is shown as a topographic (contour) map in the bottom diagram. Contour lines (brown) on the map are drawn at intervals of 20 feet, starting with 0 at mean sea level. The fact that contours bend upstream where they cross streams allows quick recognition of hilltops. Source: U.S. Geological Survey.

FIGURE 6.3

A normal closed contour (above left) encircles a small hill top. If this small hill is on the side of a larger hill, the elevation of the closed contour is the same as the higher contour, as shown here.

A depression contour (above right) encircles a pit or depression in the landscape. If the depression occurs on a slope, as shown, the depression contour has the same elevation as the lower contour. With a *C.I.* of 10 feet, the elevation of the inner depression contour is 90 feet. The bottom of the depression is less than 90 feet and more than 80 feet (because there is no 80-foot contour).

and sea level, which by definition has an elevation of zero. Figure 6.2 shows an area along a sea coast, with the sea at its average elevation of zero feet. Because the edge of the shore is everywhere at an elevation of zero feet, the shoreline coincides with the zero foot contour. If sea level rose by 100 feet, the shoreline would everywhere coincide with the 100-foot contour shown in Figure 6.2; if it rose 200 feet, it would coincide with the 200-foot contour.

Contour Interval

Contour lines are drawn on a map at evenly spaced intervals of elevation. The difference in elevation between two consecutive contours on the same slope is called the **contour interval** (C.I.). It is a constant for a given map, unless otherwise stated, and is usually given at the bottom of the map.

The choice of contour interval depends on the level of detail the topographer wishes to show and the range of elevation, or **relief,** of the mapped area. Florida is so flat that a 5-foot contour interval often best captures the landscape. The Rocky Mountains, on the other hand, show up best with 100-foot contours. A 5-foot interval would paint a Rockies map solid brown with over-abundant contours!

Index Contours

As a general rule, every fifth contour starting from sea level is an **index contour.** These are drawn as heavy lines and labeled with their elevations (Fig. 6.2). They make it easier to read a topographic map. Contours between index contours are usually not labeled.

Depression Contours

Depression contours are closed contours with hachures (short lines perpendicular to the contour line) pointing toward the lower elevations within a depression (Fig. 6.3).

They generally encircle *small* depressions, but can be used for large depressions (e.g., Fig. 6.1).

Contour Line Characteristics

The construction and reading of contour maps are governed by the following characteristics of contour lines (most of which are illustrated in Figure 6.2):

1. Every point on the same contour line has the same elevation.

2. A contour line always rejoins or closes upon itself to form a loop. This may occur outside the map area. Thus, if you walked along a contour, you would eventually get back to your starting point.

3. Contour lines never merge, split, or cross one another. However, if there is a steep cliff, they may appear to overlap because they are superimposed on one another.

4. Slopes rise or descend at right angles to any contour line.

   - Closely spaced contours indicate a steep slope.
   - Widely spaced contours indicate a gentle slope.
   - Evenly spaced contours indicate a uniform slope.
   - Unevenly spaced contours indicate a variable or irregular slope.

5. Contours usually encircle a hilltop. If the hill falls within the map area, the high point will be inside the innermost contour (however, see discussion of depression contours).

6. Contour lines near ridge tops or valley bottoms always occur in pairs having the same elevation on either side of the ridge or valley.

7. <u>Contours always bend upstream when they cross valleys.</u> Because water runs downhill, this fact allows the rapid recognition of high and low areas on a contour map.

8. If two adjacent contour lines have the same elevation, a change in slope occurs between them. For example, adjacent contours with the same elevation would be found on both sides of a valley bottom or ridge top.

9. Depression contours have the same elevations as the normal (unhachured) contours immediately downhill (Fig. 6.3).

Reading Elevations

Start with a labeled index contour. As you move uphill from this contour, keep track of the elevation by adding the value of the contour interval for every contour crossed. In Figure 6.4, moving from the 200′ index contour to point X crosses two contours: $200′ + 20′ + 20′ = 240′$ elevation. When hiking downhill you subtract contour intervals.

The elevation of a point that does not fall on a contour must be estimated. An estimate can be made by interpolation, assuming the slope between adjacent contours is uniform. For example, a point one-quarter of the way between contours with elevations of 200 and 220 feet (*C.I.* = 20 feet) would have an elevation of about 205 feet. However, slopes are often not

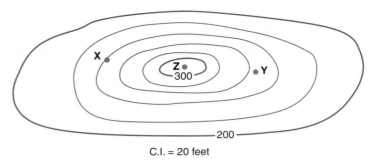

C.I. = 20 feet

FIGURE 6.4
Reading elevations from a contour map with a contour interval of 20 feet. The elevation of X is 240 feet, because X falls on a contour with that elevation. Point Y falls between the 240- and 260-foot contours, so its elevation must be between those values. Its horizontal position is about three-quarters of the way between the two, so assuming a uniform slope gives an estimated elevation of 255 feet. Point Y has a *halfway elevation* of 250 ± 10 feet: 250 is halfway between 240 and 260, and ± 10 indicates that Y falls between 240 (250 − 10) and 260 (250 + 10). Note that the error term (± 10) is found by dividing the contour interval by two. What is the halfway elevation of Z at the top of the hill? *(Answer: 310 ± 10 feet)*

uniform, so another approach is to give the *halfway elevation* between the two contours. A **halfway elevation** is the elevation halfway between the values of adjacent contours; thus, the elevation of a point between contours can be stated as the halfway elevation plus or minus one-half the contour interval. Figure 6.4 provides examples.

Study of Figure 6.3 shows that a normal closed contour that lies between a higher and a lower contour always takes the same elevation as the higher one. A depression contour in the same situation always takes the same elevation as the lower one.

Digital Elevation Models

A **digital elevation model (DEM)** consists of a high-resolution grid of points assigned elevations and colored according to elevation (Fig. 6.1B). Most DEMs are compiled from existing topographic maps. However, radar data from the Space Shuttle (SRTM), specially commissioned aircraft flights, and data from various satellites are processed to provide higher-resolution DEMs than are otherwise available from such government agencies as the U.S. Geological Survey, the Centre for Topographic Information (Natural Resources Canada), and INEGI in Mexico.

DEMs make it easier to visualize landscapes, and they often highlight subtle features that are not obvious on topographic maps. However, unless you have a computer handy, topographic maps are more useful in the field because it is easier to read accurate elevations, spot places that are easier or more challenging to hike over, and find such human-designed **cultural features** as roads, buildings, dams, and political boundaries. If you have access to Geographic Information System (GIS) software, you can drape (superimpose) a variety of topographic map features over your DEM to get the best of both worlds.

Working with Maps

We have to cover a few "necessary evils" before we can dive into landscapes and topographic maps. *Coordinate systems* are important because they allow us to precisely locate points on the Earth's surface. We also must understand the *scale* of a map so we can tell how big things are. For example, the scale of the map in Figure 6.1A tells us that Meteor Crater is about 1¼ miles across and not 50 miles across. Coordinate systems and scale are not difficult to understand once you get used to them, but they will take some extra concentration as you read the next few sections. Understanding these concepts is also important if you plan on taking a GIS class in the future.

Map Coordinates and Land Subdivision

Coordinate systems provide a permanent way of describing locations. For example, older descriptions of mineral or fossil sites commonly refer to landmarks. Unfortunately, some of these old sites are now lost because road intersections, houses, small bridges, old trees, and railway lines have since been moved or removed due to ongoing development. A coordinate system allows a state to efficiently and permanently keep track of the locations of abandoned oil wells, toxic waste sites, sealed mine shafts, and places hosting endangered plants or breeding pairs. It allows geologists to describe important rock localities, and it allows hikers to precisely locate trailheads, remote camp sites, and other places worth remembering.

Latitude-Longitude System

The most well-known global coordinate system is based on east-west lines called lines of latitude and north-south lines called lines of longitude.

Latitude measures distance north or south of the equator. The lines of latitude, also called **parallels,** form a series of parallel circles running east-west (horizontally) around the globe. The equator represents the 0° latitude line. Other parallels are set at angular intervals measured north or south of the equator, as shown in Figure 6.5A. A latitude line 40° north of the equator is termed 40° N. The **geographic poles** are at 90° N and 90° S.

Longitude measures distance east or west of the Prime Meridian. Lines of longitude, also termed **meridians,** form a series of circles running north-south (vertically) and intersecting at the geographic poles. The **Prime Meridian** is the north-south line passing through the Royal Observatory in Greenwich, England; it is defined as 0° longitude. The other meridians are set at angular intervals east or west of the Prime Meridian, as shown in

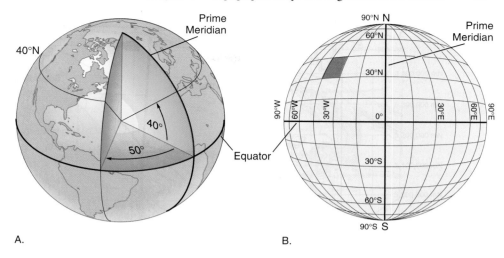

A.

B.

FIGURE 6.5

A. Cutaway view of the Earth. The east-west line 40° north of the equator is latitude 40° N; the north-south line 50° west of the prime meridian is longitude 50° W. B. Lines of latitude parallel the Earth's equator, while lines of longitude intersect at north and south geographic poles. The shaded area, bounded by lines of latitude and longitude, is a quadrangle.

Figure 6.5A. A longitude line 50° west of the Prime Meridian is termed 50° W; one 90° east of the Prime Meridian, or one-quarter the way around the globe, is 90° E. The east and west meridians meet at 180° on the opposite side of the planet. The 180° meridian corresponds to the International Date Line, except for where the date line shifts to avoid causing time-zone problems in islands or island groups cut by the 180° line. The latitude and longitude lines make a single grid network that covers the entire globe (Fig. 6.5B).

The measurement of angles from the equator (latitude) and from the Prime Meridian (longitude) is most commonly done in degrees (symbol °). A circle is divided into 360°. One degree is divided into 60 minutes (60'), and each minute into 60 seconds (60"). Maps also may indicate angular measurements in mils (one mil = 1/6400 of 360°, or 0.05625°). When measured on the Earth's surface, one degree of latitude (as measured along a meridian) is approximately 111 km (69 miles), and one degree of longitude (as measured along a parallel) varies from about 111 km at the equator to 0 km at the poles, where the meridians intersect.

A point on a map can be located by referring to the latitude and longitude of the point (for example, latitude 43° 5'

30"N, longitude 132° 15' 45"W). By convention, latitude (north-south) is given first, longitude (east-west) second.

Minutes and seconds are often expressed as decimal equivalents of a degree on global positioning system (GPS) receivers and in computer applications. A latitude of 43° 5' 8", for example, can be converted to a decimal as follows: 5' = 5/60th of a degree or 0.083°, and 8" = 8/3600th of a degree = 0.002° (there are 60 × 60 = 3600 seconds in a degree). Thus, 43° 5' 8" = 43° + 0.083° + 0.002° = 43.085°.

The U.S. Geological Survey (USGS), Centre for Topographic Information (Canada), and INEGI (Mexico) have made most of the accurate maps of their respective countries. These maps are commonly bounded by latitudes and longitudes, both of which are usually separated by intervals of 1°, ½° (= 30'), ¼° (15'), or (in the United States) ⅛° (7½'). Such maps cover rectangular-shaped areas called **quadrangles** (Fig. 6.5B), which are generally named after the largest town or most prominent geographic feature in the area (for example, the Mt. Shasta, California 7½' Quadrangle). The numerical values of the latitudinal and longitudinal boundaries are given on each map corner; intermediate values are indicated on the map margins (Fig. 6.6).

c.   What is the northern boundary latitude?   Western boundary longitude?

Southern boundary latitude?   Eastern boundary longitude?

Subtract these latitude and longitude numbers to get the size of the quadrangle in units of degrees, minutes, and seconds.

d.   What is the fractional scale of the map?

Determine the approximate verbal scale: 1 inch = _____ miles. As always, show your calculations.

An environmental restoration project requires that you enlarge part of the map to a scale of 1 inch to 1000 feet. Calculate the factor by which it needs to be enlarged.

What would the enlargement factor be if you needed a scale of 1 cm to 100 m? Hint: Start with the fractional scale.

e.   What is the contour interval?

f.   What is the highest elevation within the area designated by your instructor?

What is the lowest elevation in that area?

What is the relief of the designated area?

g.   What is the *height* (not the elevation) of the location designated by your instructor?

h.   Give the *elevation* of the location designated by your instructor.

i.   Determine to the nearest minute the approximate latitude and longitude of the designated feature.

j.   Determine to the nearest 100 m the full UTM coordinates of the designated feature.

k.   If the map is subdivided by the Township-Range method, locate the feature designated by your instructor to the nearest ¹⁄₁₆th of a section.

l.   What is the approximate size of the area designated by your instructor (in acres, if subdivided by the Township-Range method, in square meters if the UTM method is preferred)?

m.   Use the graphic scale to determine the distance in miles and kilometers between the features designated by your instructor.

n.   In what direction does the water flow in the stream designated by your instructor?

o.   What is the magnetic declination (in degrees) indicated on the map? In which year was this value measured?

2. **Analyze a landscape:** Let's say that you're a developer with a big project in mind for an area near Averill, Vermont (Fig. 6.15). You first need to study a topographic map to understand the landscape. Show any calculations you make for the following questions. Conversion factors are listed inside the back cover.

   a. Determine the following basic facts about the map:

   Interval between index contours:

   Contour interval (units in feet):

   Fractional scale (Hint: Use the UTM grid and metric units.):

   Verbal scale (1 inch = _____ miles):

   Approximate height and width of the map. (Hint: Use the UTM grid as a bar scale.)

   Approximate height and width of the map in miles (Hint: Use the verbal scale and a ruler.):

   By what factor was this map enlarged or reduced from its original 1:24,000 scale?

   b. To get a feel for the landscape, find the three most prominent mountains rising above 2200 feet. List their elevations starting with the mountain near the top of the map and going clockwise. The "T" following the bench mark elevations means they were determined from air photo measurements, which have errors of a few feet relative to the more accurate method of surveying.

   c. Determine which way the streams flow by looking at how the contours are deflected as they cross them. Draw arrows showing the flow directions of the streams flowing into or out of the ponds and lake.

   Which ponds or lakes flow into each other? (You can double-check your inferences by noting water level [WL] elevations.)

   Use the stream drainages to help you find the lowest elevation on the map. What is this elevation?

   What is the total relief of the map area?

   Let's say that you plan to hike up Brousseau Mountain from a canoe beached on Little Averill Pond. What is the height of Brousseau Mountain relative to this starting point?

   d. At the top of Brousseau Mountain you plan on checking your hand-held Global Positioning System (GPS) unit to be sure it works. Note that UGSG maps show latitude and longitude divisions no finer than 2′ 30″ (see left map margin), so you have to switch your GPS unit to UTM coordinates. Use the map to determine the UTM grid coordinates you expect to see when you reach the peak of the mountain (marked with an elevation on the map).

   e. Because your development plans include golf courses, a water park, factory outlet shopping, and extreme paintball, you need quite a bit of land around the Averill ponds. Do the little black dots on the map represent anything relevant to your development plans? Explain.

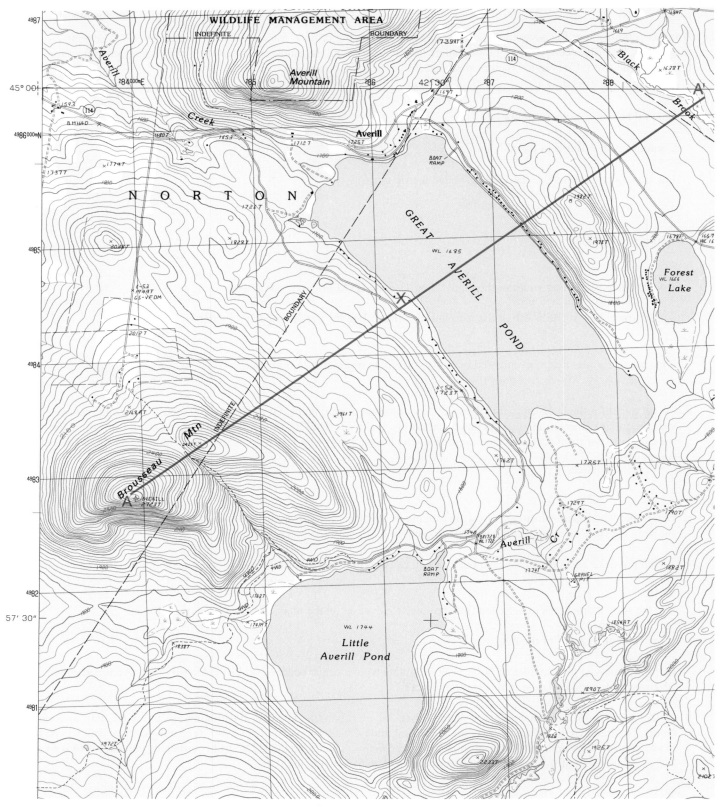

FIGURE 6.15
Portion of the Averill, Vermont, 7½-minute quadrangle
map for use in Problem 2. Canada is just a few km north
of the map area. Scale and contour interval are determined
as part of Problem 2.

110

FIGURE 6.16
Blank graph for constructing the topographic profile of Problem 2. The vertical axis marks feet above sea level.

f. Being a developer of taste and refinement, you'd like to put your name in 20-foot-tall neon letters on the top of Brousseau Mountain. But will your guests be able to *see* your name from the lodge dining area to be located at point X on the map? To find out, construct a topographic profile along the line A-A′ on Figure 6.15. To save time, use just the index contours *except* when marking the elevations of hilltops and valley bottoms. Draw your profile on the graph provided (Fig. 6.16). Label "Brousseau Mountain," "Great Averill Pond," and "Black Brook" on your profile. Draw a 20-foot letter on Brousseu Mountain and see if there is a direct line of sight from point X (the future dining room) to the letter.

Will the guests be able to see your name in lights?

What is the vertical exaggeration on the profile you drew? Show your work.

3. **Comparing a contour map with a DEM:** Figures 6.17 and 6.18 show the area around Mono Lake, CA. Use these figures to answer the questions that follow.

   a. Determine some basic facts about the map (Fig. 6.17):

   The contour interval is 200 feet. In low-relief areas, such as in Mono Valley, they have inserted supplementary contours (dashed). What is the elevation difference between a supplementary contour and an adjacent regular contour?

   Older USGS maps often emphasize the Township and Range grid system; newer maps often emphasize the UTM grid. What is the name for the areas outlined by the red squares, which are marked by such labels as R27E and T3N?

   About how many miles separate adjacent red lines on this map?

   Maps of western states frequently show many mines (most are small and abandoned) and many springs. Draw the symbols for mines and springs as shown on this map:

   Why might mappers of western states be concerned with showing every spring they find?

   b. Because 1:250,000 maps cover a lot of area, their contours tend to show only larger features. The USGS sheets also tend to be cluttered and difficult to read. In contrast, the DEM of Figure 6.18 clearly shows even subtle landscape features. The DEM image was compiled from a series of DEMs derived from the standard USGS 7½-minute topographic quadrangle maps. Comparison of Figures 6.17 and 6.18 makes obvious two advantages of DEMs: They are free of non-landscape

FIGURE 6.17

Portion of the Walker Lake (north half) and Mariposa (south half), CA, 1° by 2° quadrangles for use in Problem 3.

Original scale 1:250,000
C.I. 200 feet

clutter and, because they are based on the highest resolution maps available, they can show both broad features and fine detail, even when covering a large area. Use Figures 6.17 and 6.18 to answer the following questions:

Only fresh water flows into Mono Lake, but Mono Lake itself is very salty. Why is this? Hint: Do you see any stream leaving Mono Lake?

Since 1850, lake levels have fluctuated between 6428 feet (1919) and 6372 feet (1982). Has Black Point been an island at any time since 1850?

FIGURE 6.18

A digital elevation model (DEM) of the area around Mono Lake, CA. Lowest elevations are deep green, highest elevations are yellow. The lake level is set at 6382 feet, which is typical for the years 2000 to 2004.

Original scale 1:250,000

Lake levels have always fluctuated naturally, but from 1941 to 1982 the lake levels consistently dropped as the thirsty city of Los Angeles siphoned off more and more water from the mountain streams that feed the lake. As the supply of fresh water was cut off, how do you think the concentration of salt in the lake changed? Explain.

Los Angeles is now restricted in how much water it takes in order to preserve one of the most productive ecosystems in the world. A host of invertebrates in the lake feeds 84 different species of water birds, including 50,000 nesting California gulls!

c. Can you see anything in the DEM suggesting that lake levels were once, before 1850, considerably higher than they are today? Describe what you see and why your observations seem to be connected to lake level.

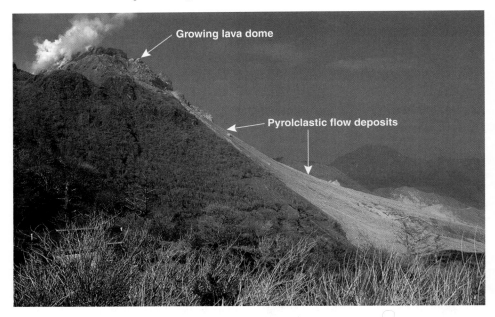

FIGURE 7.11

Mount Unzen in Japan is capped with a **lava dome** (light-colored rock under the steam). The molten interior of an active dome like this one is revealed by its dull-red glow at night. This dome is slowly growing up and out as viscous, volatile-poor lava slowly pushes out the vent. As it grows, the solid surface expands and fractures. This is why the lava dome looks like a pile of rubble. When the dome grows too tall, it collapses and sends a pyroclastic flow of partially molten blocks rolling down the hill. As the blocks disintegrate, the pressure is released on the still-molten interiors, and this allows bubbles to rapidly form, expand, and burst. This creates a cloud of scalding-hot ash like the pyroclastic flow shown in Figure 7.10. This photo was taken in 1995.

FIGURE 7.12

This ash cloud rising over Redoubt Volcano in Alaska was carried upward by hot, buoyant air heated by the ash. Note that there is a small, white steam cloud coming from the summit of the volcano. The main ash cloud is actually rising from a pyroclastic flow behind the volcano that formed when a lava dome near the summit collapsed and started tumbling downhill. As the falling dome material fragmented, it released the pressure on the still-molten lava. The pressure release allowed bubbles to rapidly form and burst. This explosive activity within the tumbling debris gave rise to the hot ash that produced this towering column of ash. Redoubt Volcano is 3100 m tall.

5 meters of mud that had the consistency and color of wet cement.

Why Volcanoes Explode

Fortunately, not all hazards are associated with all volcanoes. To further assess the hazards associated with a particular volcano, we need to understand why some volcanoes produce gentle rivers of lava whereas others explode catastrophically.

One critical factor that determines whether a volcano explodes is the concentration of volatiles in the magma. **Volatiles** include all components in a magma or lava that form gases at surface pressures and eruption temperatures. Water vapor is normally the most abundant, but other important volatiles include carbon dioxide (CO_2), sulfur dioxide (SO_2), fluorine (F), and chlorine (Cl).

At the high pressures found deep below the surface, all volatiles stay dissolved in a magma. As the magma rises toward the surface, the pressure on it decreases. At or near the surface, the pressure has dropped enough to allow the volatiles to suddenly **exsolve** (come out of solution) and form bubbles.

Since bubbles take up a lot of extra space, their sudden formation forces the volume of the magma to increase dramatically. This expansion forcefully propels the magma up and out of the volcanic vent. It is like when you open a shaken bottle of carbonated beverage: bubbles instantly form and force the beverage up and out of the neck of the bottle.

A magma low in volatiles produces a gentle eruption dominated by lava flows (Figs. 7.6, 7.7, and 7.9). However, a magma rich in gases can erupt with awesome destructive force. The rapid formation of bubbles causes such an extreme increase in volume that the volcano erupts explosively.

Viscosity is also important in determining how explosive an eruption will be. Once bubbles form in a low-viscosity magma, they rise through the fluid melt, expand in size, and merge to form still larger bubbles. When these large bubbles reach the surface, they pop with relatively little force, and, like a bubbling pot of thick, hot cereal, the popping bubbles fling relatively large droplets of melt into the air.

When bubbles form in a high-viscosity magma, they are not able to expand after they form, they cannot rise through the viscous melt, and they do not merge to form larger bubbles. They stay small and separated. As pressure decreases around the rising magma, the viscous magma keeps the small bubbles at the high pressures found at the depths where they first formed. At or just below the surface, the intense pressure inside the bubbles finally overcomes the strength of the magma, and innumerable tiny bubbles all suddenly expand with violent force. This can cause an explosion that hurls many thousands of tons of rock several kilometers through the air. In fact, these expanding bubbles can cause eruptions that are many times more powerful than the biggest nuclear bombs.

The rupturing of millions of tiny bubbles has a second important effect. When a tiny bubble pops, the walls of the bubble cool rapidly to form glassy volcanic ash. An explosive eruption rapidly distributes this ash throughout a large volume of air over the volcano. Because the ash is very hot and finely dispersed, it efficiently and rapidly heats the air. This heated air quickly expands, and this causes it to rise. This rush of rising air carries with it a great deal of ash, and this produces the characteristic column of volcanic ash that can rise 5 to 80 km above a volcano (Fig. 7.12).

In contrast, the popping of large bubbles in low-viscosity magmas flings relatively small numbers of large blobs of magma. These heat the air less effectively and produce little ash. Thus, no towering column of ash is produced over basaltic eruptions.

In sum, the eruption of volatile-poor magma of all compositions (basalt to rhyolite) produces lava flows whose length depends mainly on lava viscosity and erupted volume. For volatile-rich magmas, viscosity is critical. Low-viscosity, volatile-rich basaltic eruptions tend to produce small fountains of lava that build small cinder cones (Fig. 7.5). Eruptions of high-viscosity, volatile-rich andesite, dacite, or rhyolite can produce explosive eruptions, towering columns of ash, pyroclastic flows (Fig. 7.10), and extensive air-fall deposits. Unfortunately, the composition, volatile content, and volume of lavas erupting from even a single strato-

volcano can vary considerably over time. Mount St. Helens, for example, has erupted basalt, andesite, dacite, and rhyolite, and it has produced both lava flows (volatile-poor) and explosive pyroclastic eruptions (volatile-rich) (Fig. 7.7). Thus, while remote sensing allows us to rapidly understand the general hazards of a particular volcano, extensive on-the-ground studies are needed to comprehensively understand a given volcano.

Further Clues to Potential Violence

The high viscosity and volatile content of many of the magmas that create stratovolcanoes make these volcanoes exceptionally dangerous. They can potentially generate deadly debris avalanches, pyroclastic flows, lahars, and air-fall deposits. They can even destroy themselves in huge caldera-forming eruptions. This final section highlights a few specific features related to hazards assessment.

Sector Collapse

A rising body of magma may cause a volcano to bulge out one side until part of the peak (a sector) collapses and generates a debris avalanche. This **sector collapse** leaves behind a horseshoe-shaped ridge (Fig. 7.13), not the large, full ring of a perfect caldera (Fig. 7.8). The collapse may trigger a major eruption by releasing pressure on the underlying magma and thereby allowing explosive bubble formation.

A second, more subtle, process may also lead to sector collapse. Low levels of igneous activity can fuel persistent circulation of hot water and steam within a volcano. These hot, acidic fluids alter and weaken rock within the volcano. A minor eruption, earthquake, or even a violent rainstorm can trigger sector collapse and debris avalanches. The steep, upper slopes of Mount Rainier in Washington have been weakened in this way.

Lahars

A volcano capped with snow and ice or a crater lake is at substantial risk of producing a lahar. Even minor eruptions, which are far more common than major eruptions, can produce enough melting to generate these deadly mudflows. The huge mass of glacier ice on Mount Rainier makes lahars

FIGURE 7.13
The May 18, 1980, eruption of Mount St. Helens was triggered by a sector collapse: A growing bulge on the north side collapsed, which caused the suddenly exposed magma to explode. As the eruption progressed, the crater expanded as more rock fell into it and was blown out by the on-going explosive activity. The result is a crater opening out in the direction of the collapse. A caldera forms a much larger feature as much of the volcano collapses into the partially emptied magma chamber (see Fig. 7.8). The tiny bump in the center of the crater is a growing **lava dome.** In 1983, when the elevation data for this image were collected, the dome was about 100 feet above the crater floor. By April 2006 it had grown to over 400 feet above the crater floor.

from this volcano the biggest hazard in the Cascades (Fig. 7.14). To spot lahar hazards, look for the white or bluish colors of snow and ice on air photos and satellite images of volcanic peaks.

Lava Domes

Lava domes may dominate the summit of a volcano (Fig. 7.11) or they may appear small compared to the rest of the volcano (note the very small dome within the caldera in Figure 7.13). Lava domes that are growing near the summit of a volcano are in an excellent position to collapse with little warning and thereby produce a deadly pyroclastic flow that sweeps across communities living on the lower slopes of a volcano (Fig. 7.11).

Erosion

Erosion is an important clue that a given volcano may be inactive. Although all tall mountains experience rapid erosion,

FIGURE 7.15
Stereographic pair of one of the Menan Buttes near Idaho Falls, Idaho. These tuff cones are mid- to late-Pleistocene in age.

g.   Examine Figure 7.15 using a stereoscope provided by your instructor. Would you say that this volcanic cone is steep-sided or broad and flat? Is the rim of the crater generally sharp or is it rounded and broad?

h.   Recall that air photos make objects appear three or four times taller than they are. To get a more accurate view of the relief of the Menan Buttes, make a topographic profile on the graph below following the line A-A′ in Figure 7.16. Use the 100′ elevation increments of the index contours, and be sure to locate the tops of ridges and the bottom of the crater.

i.   How wide is this tuff cone at its base (in miles)? How tall is it (in feet)? Show your work.

j.   Given their size, height, and basaltic composition, do you think that tuff cones such as these represent a substantial volcanic hazard? Are they likely to produce vast lava flows, scalding pyroclastic flows that travel 10 or 20 miles from the vent, roof-crushing air-fall deposits, or devastating lahars? Why?

Scale 1:24,000
C.I. 10 feet

FIGURE 7.16
Portion of the 7½ minute Menan Buttes, Idaho, topographic map.

2. Science advances when scientists' general expectations of how nature works are confounded by careful observations. The new observations force us to refine and improve our understanding. You probably grew up assuming that volcanoes normally erupt from their summits. We will see if this is true of the volcanoes that built the big island of Hawaii by examining a digital elevation model (DEM; Fig. 7.17) and satellite image (Fig. 7.18) of the island.

    a. All of Hawaii was built from the eruption of lava from volcanoes. Examine Figures 7.17 and 7.18 to locate the major volcanic centers that built (and are building) this island. Circle the location of each on the DEM (Fig. 7.17). How many <u>major</u> volcanic centers are there? Note: One center does not show up well in terms of topography. Look for the dark lava flows in Figure 7.18.

    b. Rock weathering and soil formation progressively erase old lava flows. Use this fact to determine the relative age of each volcanic center. On Figure 7.17, number your circled volcanoes in order of age with "1" being the oldest. Hint: The youngest volcanic center is issuing thin clouds of steam (Fig. 7.18).

    c. Closely examine Mauna Loa (Fig. 7.18) to see where most of the young (black) lava flows originate. Do most come from the summit caldera, volcanic cones on the flanks, or from one or more fissures extending away from the summit?

    d. The topographic profile below shows both the gentlest and steepest slopes of Mauna Loa. The letters below the profile refer to the bends in the line segment in Figure 7.17. Segment A-B-C profiles the gentle south slope of Mauna Loa whereas segment C-D shows the steeper southeast slope. Measure the width of the profile at sea level, the height, and the average slopes. Show any calculations.

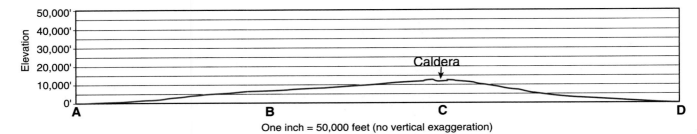

One inch = 50,000 feet (no vertical exaggeration)

    e. Measure the length of the lava flow marked "A" on Figure 7.18. You will first need to determine the scale of the image using the fact that the distance between the two east-west UTM grid lines is 100 km (62 miles). Show your work.

FIGURE 7.17

Digital elevation model (DEM) of the big island of Hawaii. The relative size of large shield volcanoes and typical cinder cones is illustrated by the fact that some of the tiniest bumps on this image are cinder cones formed on the flanks of the shield volcanoes. The lines A-B-C-D follow the profile used in Problem 2d. North is to the top of the image. Scale = 1:1,150,000.

FIGURE 7.18

False-color Landsat 7 satellite image of the big island of Hawaii. Black and gray represent bare volcanic rock, tan and rusty orange represent volcanic soils and sparse uplands scrub vegetation, and green represents tropical vegetation. Faint plumes of steam rising from two vents are bluish white, the ocean is blue, and clouds are white. The thin-black lines in the ocean are part of the UTM grid; those on land are roads and, at Hilo, a breakwater. Two east-west UTM lines 100 km apart are labeled. "A" marks a lava flow for use in Problem 2e. North is to the top of the figure.

3. The eruption of Mount St. Helens on May 18, 1980, was the deadliest, most destructive, and most explosive in the recorded history of the lower 48 states. It began when rising magma inside the volcano forced a large bulge to grow out its north side. A small earthquake caused the bulge to collapse, and this created a massive debris avalanche. This sector collapse also released the pressure on the volatile-rich, dacitic magma rising within the volcano. Bubbles formed so suddenly and with such force that the volcano directed a powerful lateral (sideways) blast that felled an old-growth forest over a broad area. As the eruption continued, the volcano started blasting material vertically and a towering column of ash 25 km tall (as in Figure 7.12) was established. Fifty-seven people, including one young volcanologist monitoring the volcano, were killed by this eruption.

   a. Comparison between Figures 7.19 and 7.20 shows the area stripped of vegetation by the debris avalanche and lateral blast. How many kilometers out from the central crater were deforested by the blast? Figure 7.20 covers a north-south distance of 48 km. Measure the distance from the crater to the edge of the blast zone north of Spirit Lake. Show your work.

   b. Comparison of the two satellite images also reveals that a number of stream valleys were filled by light-gray material by the eruption. Do you think that this material was deposited by lava flows, pyroclastic flows, ash fall, or lahars? Explain your choice. Hint: Note that the pre-eruption peak of Mount St. Helens was encased in snow and glaciers (Fig. 7.19).

   c. The debris avalanche sent the largest landslide in recorded history down the north side of Mount St. Helens. Much of this material coursed down the upper reaches of the North Fork Toutle River (Fig. 7.20). Compare the number of lakes in Figures 7.19 and 7.20. (Note: Spirit Lake looks like it was split in two. In fact, this "split" is just floating logs from the forest that was blasted down on May 18th. These logs are still floating; they move around depending on the wind.) How did the new lakes form?

   d. Use a stereoscope to examine the crater area in Figure 7.20. What has formed in the crater? What does the future hold for the shape of Mount St. Helens if moderate eruptions continue to build this feature over the long term?

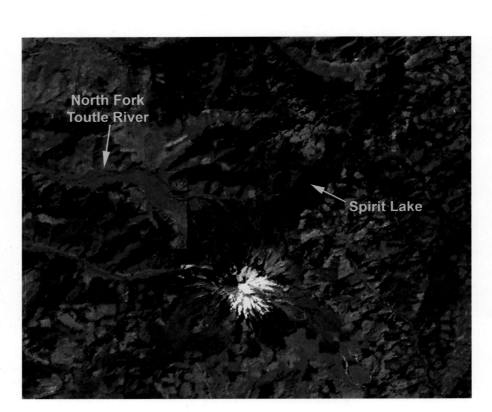

FIGURE 7.19

This false-color, relatively low-resolution Landsat 1 satellite image from 1973 captures Mount St. Helens before its 1980 eruption. The dark green indicates abundant vegetation (mainly evergreen forests) north of the snow-capped volcano. Grey indicates rock or bare soil and white is snow. The black area north and slightly to the east of the volcano is Spirit Lake; its clear waters reflected no infrared or visible light back to the satellite. Scale = 1:255,000. North is toward the top of the image.

e. Measure on the profile below the width at the base (elevation of ~4,000'), the height, and the steepest slope of the pre-eruption cone of Mount St. Helens. Show any calculations.

f. Mount St. Helens has erupted material ranging in composition from basalt to dacite. In the weeks leading up to the May 18th eruption, geologists sampled the ash ejected by small steam eruptions in hopes of discovering the composition of the rising magma. What rock composition and volatile content would have predicted a relatively gentle eruption?

FIGURE 7.20

This stereopair shows Mount St. Helens in 1992, 12 years after the major eruption of 1980. It combines a false-color Landsat 5 satellite image with elevation data from Space Shuttle Radar Topography Mission flown in 2000. Green is vegetation, gray is rock, sediment, or bare soil, and several lakes appear as black or greenish black. The image on the left covers an area of 48 × 25.5 km. North is toward the top of the image.

4. Naples and a host of smaller towns and rural communities are built around Mount Vesuvius in Italy (Fig. 7.21). In addition, Campi Flegrei, an active caldera mostly buried by a number of small volcanic centers, is visible just west of Naples. Campi Flegrei, better known in English as the Phlegrean Fields, is the site of famous Roman ruins.

    a. There is a long, arc-shaped ridge north of Vesuvius' crater that marks the remains of an old stratovolcano called Mount Somma. This volcano was largely destroyed in violent eruptions roughly 18,000 and 16,000 years ago. Vesuvius is now growing on the remains of Mount Somma and has buried most of its southern portions. Mount Somma was originally much taller than Vesuvius; where do you think much of Mount Somma went? What type of volcano did Mount Somma become following its big eruptions 18,000 and 16,000 years ago? Hint: There are no associated debris avalanche deposits like those found at Mount St. Helens.

    b. Examine the topographic profile of Vesuvius below. What type of volcano does its size and slope most closely resemble?

    Is this the type of volcano that you would expect to produce violent, explosive eruptions?

    c. The satellite image (Fig. 7.21) spans 36 km from north to south. How many kilometers are represented by one centimeter? Show your work.

    d. Vesuvius is an active volcano that has erupted many times before and after the famous blast of 79 A.D. (in which Pompeii was buried). Campi Flegrei has erupted twice in historical times. The current emergency evacuation plan assumes a worst case in which an explosive volcanic eruption lofts a column of searing-hot ash and other pyroclastics 20 km or more into the air. Pyroclastic flows will race down one or more sides of the volcano at up to 240 km/hr with a force able to flatten walls 3 meters thick! Every living thing within 7 km of the crater is at risk of being flash-fried by such a flow.
        Use a compass and the number you got in problem 4c to draw a circle with a radius of 7 km centered on the vent. Label this circle "pyroclastic flows."

    e. Vesuvius has experienced two eruptions involving a sector collapse like Mount St. Helens. To assess the potential hazard associated with a Mount St. Helens-style lateral blast, use a compass and the number you got in question 3a to draw a circle of appropriate radius around Vesuvius. (If you were not assigned problem 3a, get the radius from your instructor.) Label this circle "lateral blast." Would Naples suffer destruction due to the direct effects of a lateral blast as powerful as that of Mount St. Helens?

    f. Another caldera-forming eruption is not likely for Campi Flegrei. Nevertheless, to illustrate the power of such eruptions, assume that a second caldera 13-km in diameter is produced by the next major eruption. Use a compass to center this caldera on the heavily vegetated large crater in Campi Flegrei (Fig. 7.21). Would any of the inhabited areas around Naples fall within the caldera?

    Recent archeological excavations have found evidence for the most powerful known eruption of Mount Vesuvius. The eruption of 1780 B.C. produced lahar deposits 3 meters thick extending into the city limits of modern Naples and ash-fall deposits thick enough to collapse modern roofs over most of the Naples area. The ash deposits preserve thousands of footprints of Bronze-age people evacuating the area after the eruption.

FIGURE 7.21
This false-color ASTER satellite image of Naples, Italy, includes the infamous Mt. Vesuvius (right of center) and the Campi Flegrei (far left). The caldera rim of Campi Flegrei is obscured by younger volcanic centers on land, and is below sea level to the south. Red indicates growing vegetation, blue indicates roads and buildings, and black is the Mediterranean Sea. This image covers an area of 36 × 45 km. North is up.

Volcanoes on the Web

5. The Web is a rich resource for information on volcanoes around the world.

    a. There are two sites that feature an encyclopedic coverage of world volcanoes big and small. The Smithsonian Institution maintains a website with information related to any volcano that has erupted during the past 10,000 years: *www.volcano.si.edu* (or link to it through *www.mhhe.com/jones6e*—see Preface). The University of North Dakota also hosts an excellent website focusing on volcanoes around the world: *http://volcano.und.edu/*. Each has pages devoted to recently active volcanoes as well as entries for hundreds of volcanoes around the globe.

    Use one or both sites to find out more about two recently active volcanoes (specified by your instructor or listed on the websites). Fill out as much of the table below as you can:

Volcano Name	Location	Type of Volcano	Main Hazards

    b. The United Stated Geological Survey (USGS) operates a volcanic hazards program and five volcanic observatories whose mission is to monitor and study the nation's active and potentially active volcanoes. The USGS volcanic hazards web page (*http://volcanoes.usgs.gov/*) includes an excellent guide to volcanic hazards illustrated by real case studies (click on *Types & effects*) and links to the five volcanic observatories, which are rich in excellent photos (click on *Alaska, Cascades, Hawaii, Long Valley,* or *Yellowstone* in the bar near the top).

    Visit the appropriate USGS volcanic observatory web page to research either a volcano near you or one that your instructor specifies. Find out each of the following about your volcano:

    1. What type of volcano is it (cinder cone, shield, stratovolcano, or another type that you specify)?

    2. Is it active, dormant, or extinct?

    3. What are the main hazards represented by your volcano?

    4. Is there a volcanic hazard map? Do you or someone you know live in a hazard zone? If so, give the type(s) of hazard(s).

    c. A web site filled with great photographs, panoramas, and spectacular video clips from many interesting volcanoes is Stromboli Online: *www.swisseduc.ch/stromboli/*. When you watch the video clips, it is worth having speakers attached to your computer to get a sense of the explosive force associated with even the relatively weak popping of bubbles in basaltic lavas. Examine the images and clips to classify the volcanoes in the table below based on their approximate size and shape.

    To help with this, go to *http://volcanoes.usgs.gov/* and click on *Photo glossary* to get illustrated definitions of potentially unfamiliar terms. Keep this page open while you browse Stromboli Online.

Volcano Name	Type of Volcano	Volcano Name	Type of Volcano
Stromboli		Vanuatu (Lopevi)	
Etna		Erte Ale	
Soufrière Hills		Ol Doinyo Lengai	
New Zealand (see links to Ruapehu eruption of 1996, its lahars, and air photos)		Tenerife (see Geological Background, Pico de Teide)	

In Greater Depth

6. In this problem you will make a graph showing how the ash cloud from the May 18, 1980 eruption of Mount St. Helens moved across the country. The ash did not accumulate enough to collapse roofs, but the highly abrasive material did accumulate like a thin layer of snow over a broad area and caused problems with engines and other machinery. You can see the movement of the ash cloud in a series of images (Fig. 7.22) taken by the Geostationary Operational Environmental Satellite (GOES) from 39,500 km. The times at which the images were taken are given (Pacific Daylight Time), and the state boundaries are superimposed on the images.

   a. Do your best to trace an outline of the volcanic ash cloud on each of the GOES images. Transfer these outlines from each image onto the map (Fig. 7.23) using the state boundaries as guidelines. Label each ash-cloud boundary with the time of the image.

   b. Use Figure 7.23 to measure the *maximum* distance from Mount St. Helens to the leading edge of the ash cloud for each image. Enter your results in Table 7.1.

   c. Calculate the travel time represented in each image by subtracting the time of the eruption (08:32) from the time of each image. Enter the result in Table 7.1.

   d. Graph the data of Table 7.1 on Figure 7.24 by plotting the distance traveled, in kilometers, on the horizontal axis versus the travel time on the vertical axis for each image. Connect the data points. When the points are connected, do they form a straight line, a smooth curve, or an irregular line?

   e. The velocity at which the ash plume moved is represented by the slope of the lines connecting the points—the steeper the slope, the slower the velocity. What does the graph tell you about the wind velocity? Was it constant, did it gradually increase or decrease, or was it variable?

   f. The velocity at which the ash plume traveled can be calculated by dividing the distance traveled during a particular time interval by the time interval. For example, if the ash cloud traveled 425 km in 200 minutes, velocity, v = 425 km/200 min = 2.13 km/min. Calculate the overall, or *average*, velocity between 8:32 and 16:15. Give the answer in (1) km/min, (2) km/hour (multiply [1] by 60), and (3) miles/hour (1 km = 0.62 miles).

   g. In what general direction was the prevailing wind (see Fig. 7.22)?

      Was the direction the same along the entire path, or did it change? If so, how?

   h. Based on the shading of the ash plume during the period of the images and on the outlines you drew on the map (Fig. 7.23), predict where the thickest deposits of air-fall ash were likely to have accumulated.

FIGURE 7.22

Satellite images taken by GOES from a geostationary orbit 39,500 km above the Earth illustrate the movement of the ash cloud following the May 18, 1980, Mount St. Helens eruption. The ring in the first image is a shock wave. The ash cloud is inside the ring. As the original scientists examining these images had to do, do your best to distinguish between the expanding ash cloud and the normal water vapor clouds that were present on the day Mount St. Helens erupted. For use with Problem 6.

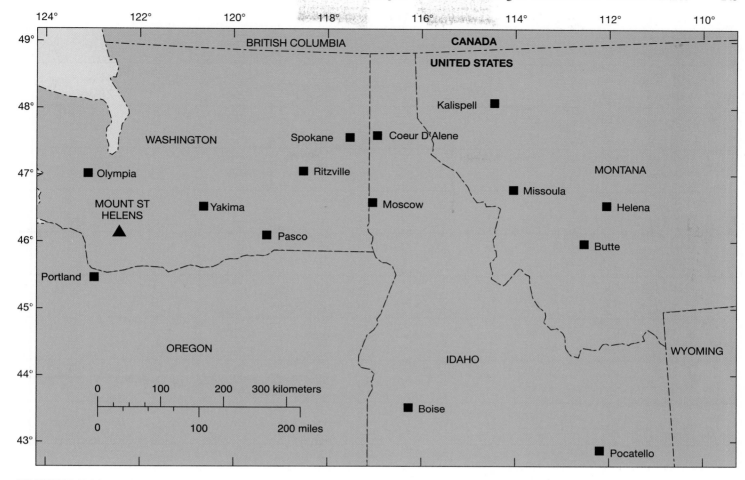

FIGURE 7.23
Outline map of area affected by ash from Mount St. Helens. For use with Problem 6.
Source: U.S. Geological Survey.

Table 7.1 GOES Data for the Eruption of Mount St. Helens, May 18, 1980		
Time of Image (hh/mm)	**Maximum Distance from Mount St. Helens**	**Travel Time (minutes)**
9:15		
10:15		
11:15		
12:15		
13:15		
14:15		
15:15		
16:15		

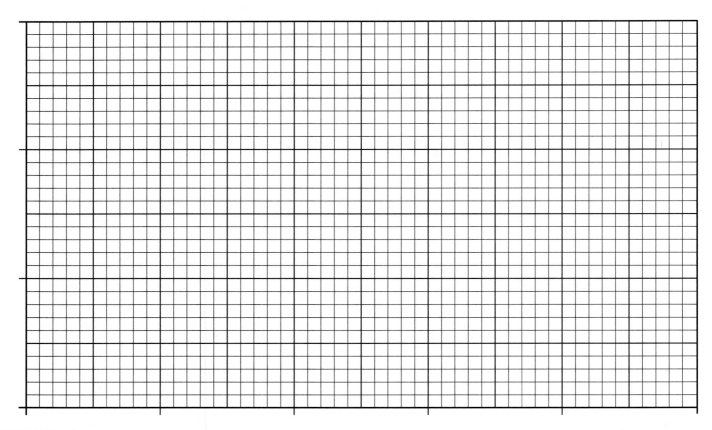

FIGURE 7.24
Graph for use with Problem 6.

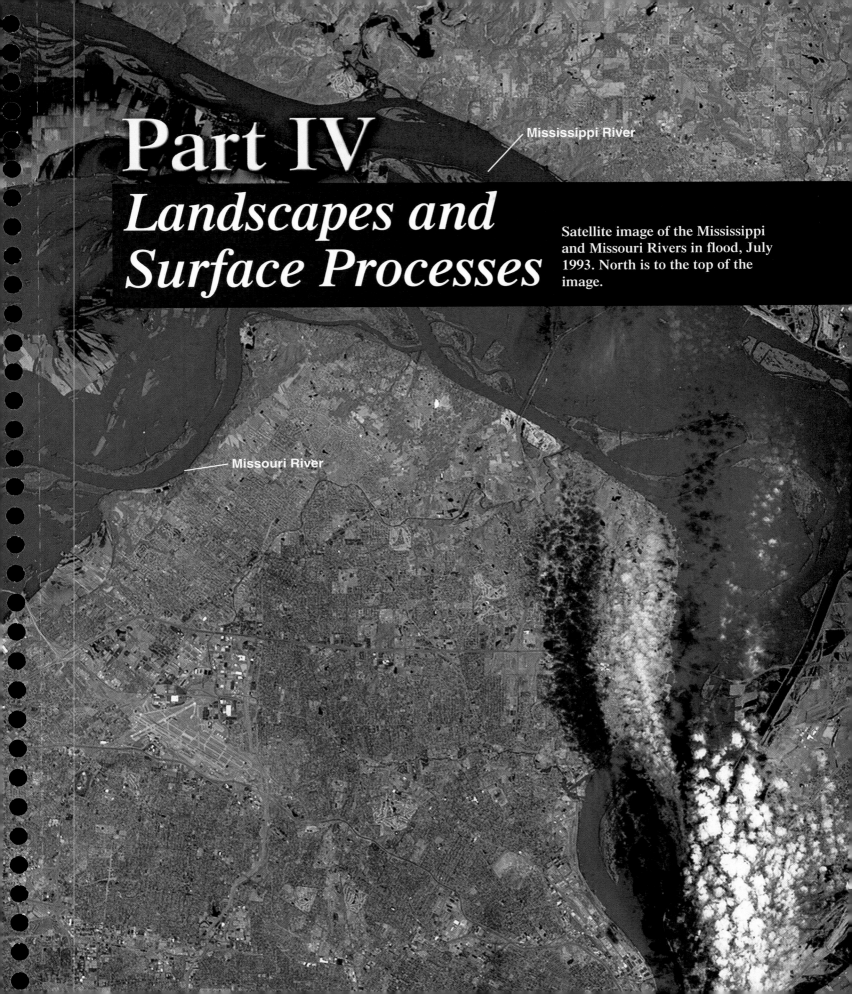

Part IV
Landscapes and Surface Processes

Mississippi River

Missouri River

Satellite image of the Mississippi and Missouri Rivers in flood, July 1993. North is to the top of the image.

CHAPTER 8

Streams and Humid-Climate Landscapes

Materials Needed

- Pencil and eraser
- Colored pencils
- Calculator
- Ruler

Introduction

Running water is the most important force shaping the Earth's surface. Even desert landscapes are dominated by drainage systems carved by infrequent rainstorms. Thus, many familiar landforms owe their origin to streams. In addition, many cities are located on rivers and depend on them for drinking water, freight transport, and recreation. In this lab you will learn how to determine the size, shape, and runoff characteristics of a drainage basin and to recognize and interpret the common erosional and depositional features formed by running water. You will see how these features depend on the underlying geology, climate, and changes in elevation relative to base level and time. You will also learn to use data from gaging stations to predict the frequency, size, and extent of major floods using data available on the World Wide Web.

This lab focuses on landscapes produced by streams in *humid* regions—those with annual precipitation of more than

50 cm (Fig. 8.1). Such regions are characterized by abundant vegetation, thick soils, and relatively rounded ridges, valleys, and slopes. Chemical weathering typically predominates over mechanical weathering, and mass wasting transports sediment downhill, where it is eventually carried off by streams.

Runoff and Drainage Basins

Streams are part of the hydrologic cycle, as shown in Figure 8.2. They carry water precipitated on the surface back to the oceans as surface runoff. Some of the precipitation returns to the atmosphere by *evaporation* and *transpiration* (emission through the leaves of plants), and some soaks into the ground and is carried away below the surface (the *infiltration loss* in the equation below). Water that enters the streams and makes up the *runoff* comes from direct overland flow and from water that moved underground before being discharged into

the streams. Surface runoff, the runoff due to overland flow, can be expressed as:

surface runoff = precipitation – infiltration loss – evaporation – transpiration.

Anything affecting one of these terms affects surface runoff, so the percentage of precipitation that naturally runs off varies considerably worldwide. In Arizona, where precipitation is about 25 cm/year, evaporation and infiltration are high, and runoff is very low, typically 2 cm/year or less. In contrast, precipitation in Alabama is about 125 cm/year, and, because infiltration and evaporation are low, runoff is about 75 cm/year. Human activity also affects runoff. For example, runoff increases when the land surface is paved so that rain doesn't soak into the ground, or when vegetation is removed so that the amount of transpiration is reduced.

When water runs off a surface, it flows downhill and into a stream. The stream is part of a drainage network in which smaller streams feed larger streams. As shown in

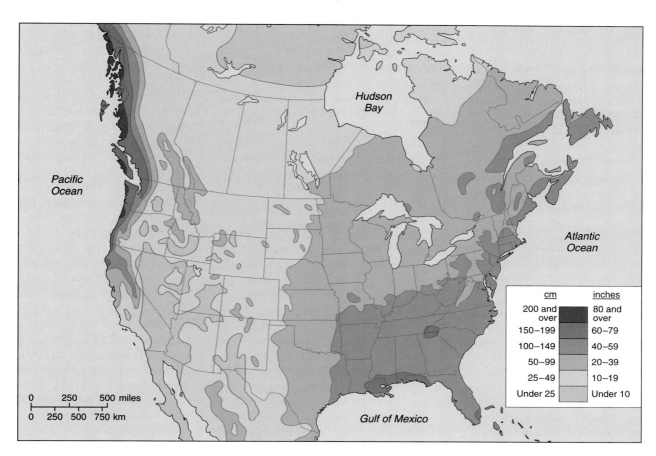

FIGURE 8.1

Average annual precipitation in the United States and Canada. Areas with more than 50 cm per year are considered humid. From *Geosystems,* 2nd edition by Robert W. Christopherson. Copyright © 1994. Reprinted by permission of Prentice-Hall, Inc. Upper Saddle River, N.J.

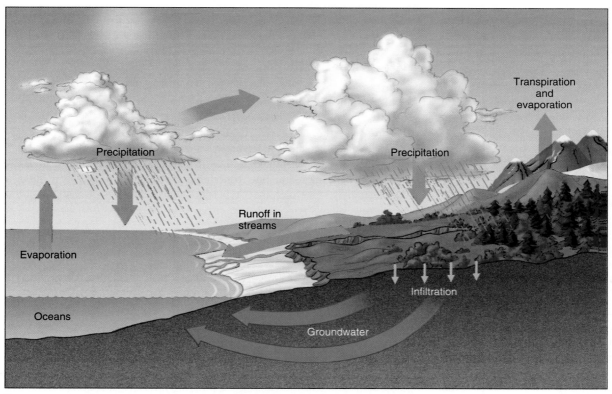

Water resources: Oceans and inland seas, 97.22%; glaciers, 2.15%; ground water, 0.62%; lakes, rivers, and streams, 0.01%

FIGURE 8.2

The hydrologic cycle and relative volumes of Earth's water resources. This chapter emphasizes runoff.

147

Figure 8.3, each stream, no matter how large or small, has an area that it drains called a **drainage basin** or **watershed.** Drainage basins are separated by **divides,** which are higher points between basins. Rain falling on one side of a divide goes downhill into one drainage basin; that falling on the other side goes into another drainage basin.

Using Clayborn Creek in Figure 8.3 as an example, divides can be located, and the drainage basin outlined, as follows:

1. Locate the stream in question (Clayborn Creek) and its tributaries (branches) and note which way each is flowing (i.e., which way is downhill). Note how contours crossing valleys "point" uphill.

2. Locate other streams on the map and note which direction they are flowing. For example, in Section 31 (red number 31 on Figure 8.3) a tributary of Clayborn Creek flows south-southwest from the vicinity of Inspiration Point. Another stream in the northwest corner of Section 31 flows north. Both streams flow away from a divide into their own drainage basins.

3. Find the highest point on a line between the ends of the two streams. That point will lie on a divide and, therefore, on the margin of the drainage basin of Clayborn Creek. In this case, the elevation is about 1440 feet.

4. The divide here is a well-defined ridge. In this case, a road more or less follows the ridge on the north side. By following the ridge, the drainage basin of Clayborn Creek can be outlined (the dashed line).

Stream Channels and Valleys

Figure 8.4 shows a contour map of a stream valley and a topographic profile drawn along its length. The **longitudinal profile** illustrates how the **gradient** (see Chapter 6) gradually decreases from the **head** to the **mouth** of the stream. The mouth is the **base level** for a stream: it limits the depth to which that stream can erode. Over time, a stream adjusts its channel and longitudinal profile in response to changes in discharge, base level, and erodibility of the rock or sediment over which it flows. Ideally, the adjustments lead to a near *balance* between erosion and deposition along the course of a

stream and produce a smooth longitudinal profile, as shown in Figure 8.4. A stream that does not have a smooth profile erodes or deposits so as to attain one: waterfalls and rapids are eroded, lakes or ponds along streams are filled.

The size of a stream channel and the volume of water increase downstream. The volume of water per unit of time is the **discharge** and is given by this equation:

$$discharge = velocity \times cross\text{-}sectional$$
$$area\ of\ channel$$

Common units for discharge are cubic meters per second (m^3/sec) or cubic feet per second (ft^3/sec); for velocity, meters per second (m/sec) or feet per second (ft/sec); and for cross-sectional area, square meters (m^2) or square feet (ft^2). As discharge increases, so do all of its components. Thus, during flooding, velocity and channel size increase as the volume of water increases. Stream velocity, channel area, and discharge are recorded at *gaging stations* on many streams throughout the United States. For example, during the summer flood of 1993 (see Part IV opening image, p. 145), the gaging station on the Mississippi River at St. Louis, Missouri, recorded a peak discharge of 1,070,000 ft^3/sec on August 1. This is more than eight times the average August discharge of 133,000 ft^3/sec.

Features of Streams and Their Valleys

Streams vary, from turbulent mountain streams rushing down narrow valleys to great rivers with wide valleys flowing across a nearly flat landscape. As streams vary, so do their characteristic features.

Streams with steep gradients tend to erode downward more rapidly than they erode laterally. Therefore, they typically have narrow valleys with V-shaped cross-

profiles. Longitudinal profiles are irregular because of the presence of waterfalls and rapids. Figure 8.5 illustrates these features.

With decreasing gradient, lateral erosion becomes more important, and broad valleys develop. Such streams have a variety of distinctive features, as illustrated in Figure 8.6. The actual stream channel is much narrower than the valley, most of which is occupied by the **floodplain,** the area that could be submerged during a flood. Just adjacent to the channel are **natural levees,** low ridges formed by sandy sediments rapidly deposited by flood waters. **Backswamps** develop in low areas on the floodplain behind natural levees. A river channel is not straight, but meanders or winds about in the floodplain; a bend in the channel is a **meander.** Erosion on the outside of a meander forms a **cutbank,** and deposition on the inside of a meander forms a **point bar.** This combined erosion and deposition causes the meanders to move across the floodplain. The **meander belt** is the zone in the floodplain within which meanders occur. The channel may take a shortcut across a meander loop to form a **cutoff,** or abandon the loop altogether to form an **oxbow lake.** Where floodplains are wide and natural levees are high, tributary streams may flow in the floodplain for long distances before joining the main river; such tributaries are known as **yazoo streams. Stream terraces** are step-like benches above the level of the present-day floodplain. They represent the remnants of preexisting floodplains or valley floors.

Floods and Recurrence Intervals

A flood occurs when a stream overflows its channel. The size of a flood, as measured by maximum discharge, or by **stage** (elevation of water surface), varies from year to

FIGURE 8.3

(Page 149) The drainage basin of Clayborn Creek and its tributaries is outlined on this portion of the Beaver, Arkansas/Missouri, quadrangle. Note that the divides connect the highest elevations between adjacent drainage basins. Water falling within the drainage basin moves downhill toward Clayborn Creek and eventually makes its way into the White River.

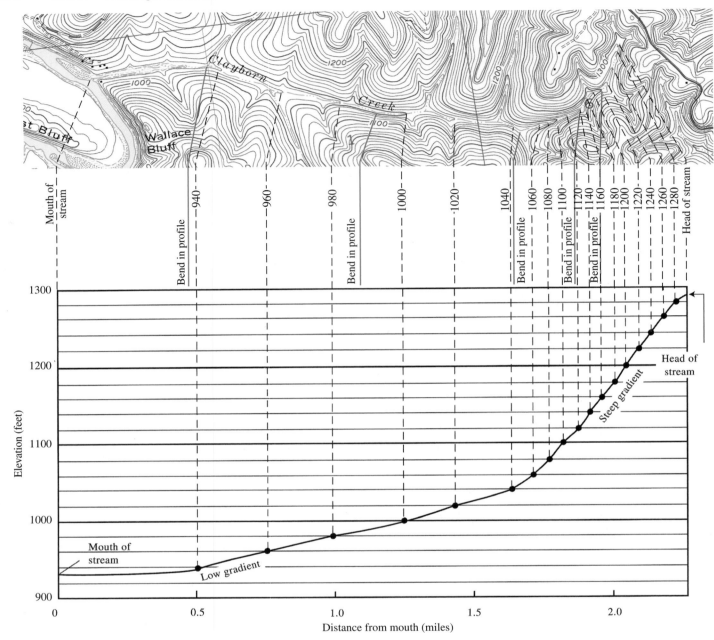

FIGURE 8.4

The longitudinal profile along Clayborn Creek was drawn by marking, on the top of the profile paper, points where contours cross the stream and labeling these points with their elevations. The profile follows the bends in the stream, not the straight-line distance between the head and the mouth. An easy way to follow along the stream when marking the contour-crossing points is to hold the paper down with a pencil where the stream bends, then rotate the paper to follow along the next stream segment. The locations of the bends are shown here for illustrative purposes only. Note that the gradient decreases from the head to the mouth of the stream. The vertical exaggeration is 15.9 times. (The horizontal scale is 1:22,850, and the vertical scale is 1 inch to 120 feet.)

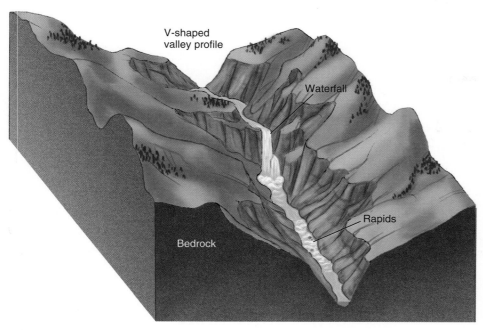

FIGURE 8.5
Features of a stream with a steep gradient.

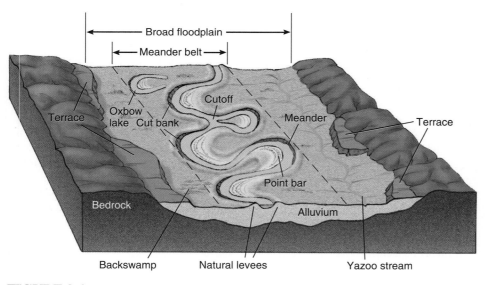

FIGURE 8.6
Features of a stream with a gentle gradient.

year. By analyzing the frequency of floods of various sizes, a *recurrence interval* can be developed for a river at a particular locality. The **recurrence interval,** usually measured in years, is the average interval between floods of a particular size. Thus, *on average,* a 100-year flood will recur at intervals of 100 years. That does not mean that a flood that size could not occur two years in a row; it means that the chance of it occurring in *any* year is 1 in 100. A *flood-frequency curve* plots discharge, or in some cases, stage, against recurrence interval, as illustrated in Figure 8.7.

Floodplain zoning is based on recurrence intervals. If planners know what stage a stream will reach in a 50-year flood, they can determine what parts of a floodplain are likely to be flooded about once every 50 years. They can then zone accordingly, deciding which areas are better for homes, businesses, and hospitals and which for parks and recreation. The width of a 50-year floodplain is determined from gaging-station records and flood-frequency curves. However, this width may be uncertain if a gaging station was in place too few years or if there is no station on the stream in question.

Landscapes in Humid Areas

Drainage Patterns

Drainage patterns reflect the characteristics of the underlying rocky materials (Fig. 8.8). Landscapes developed on flat-lying, homogeneous rock or unconsolidated sediments will naturally develop a **dendritic** (branching like a tree) drainage network. Water running off a volcano or uplifted dome will create a **radial drainage pattern** of streams flowing from a central point. If the underlying rock has been cut by fractures, these are more easily eroded and cause drainages to form a **rectangular drainage pattern.** In mountainous areas where rock layers have been folded and eroded to expose long resistant ridges, most streams will run along the valleys to join a larger river that has cut across the ridges. This forms a **trellis drainage pattern** (Fig. 8.8).

Evolution of Stream Systems

Stream valleys change with time in ways that are fairly predictable *if* the controlling factors remain constant. The following characteristics can describe the evolution of a single, long-lived drainage system or the features of a single river moving from the headwaters to the midsection to the lowlands, especially if the headwaters are in the mountains.

Early-stage streams show steep gradients down to a base level created by a lake, larger river, or ocean. High-gradient streams cut deep into the landscape and form V-shaped valleys with narrow floodplains (Fig. 8.5). Waterfalls or rapids frequently occur. The surrounding landscape shows high hills, deep valleys, and prominent drainage divides.

With time, erosion does three things: it cuts farther back into the hills in a process called **headward erosion** (Fig. 8.9); it brings the elevation of most of the stream channel nearer to base level, thus reducing

FIGURE 8.7

Flood-frequency curve for Rock Creek near Red Lodge, Montana, 1932 to 1963, using maximum annual discharge data from the USGS. From the *Recurrence Interval (RI)* scale (bottom), it can be predicted that, on average, a discharge of 2600 ft³/sec will be attained every 20 years. Using the *Percent Probability of Recurrence (P)* scale on top, the same thing can be said in a different way: there is a 5% probability that a discharge of 2600 ft³/sec will be reached in a given year. $P = 100/RI$.

the river gradient; and it reduces the elevation and relief of the surrounding landscape. As a result, **middle-stage streams** are characterized by longer drainage systems with more tributaries, moderate gradients, and increasingly rare waterfalls and rapids. In addition, lower gradients reduce the importance of downcutting while increasing the importance of lateral (side-to-side) erosion. Lateral erosion creates meandering channels, wider flood plains, and broad, flat valleys within otherwise hilly landscapes.

With still more erosion, middle-stage streams become **late-stage streams,** which feature low gradients and extensively developed meander belts characterized by wide floodplains and extensive systems of oxbow lakes, backswamps, and occasional yazoo rivers (Fig. 8.6). While low hills may bound one or both sides of a late-stage floodplain, the surrounding landscape is overall fairly flat.

Headward erosion puts streams in adjacent drainage basins in competition with each other. Clayborn Creek (Fig. 8.3), for example, is trying to expand its area while at the same time the surrounding creeks are trying to expand their areas. If erosion rates on either side of a divide are the same, headward erosion stalls, the position of the divide stays about the same, and the landscape merely loses elevation. If erosion rates are higher on one side, headward erosion causes the faster-eroding stream to cut into the drainage basin of the other stream as it wears down the landscape. Eventually, the dominant stream can snare one or more tributaries of the other stream, thus **beheading** it in a process known as **stream capture** or **stream piracy** (Fig. 8.9).

Early geologists formulated the idea of stages in stream development in terms of a grand *erosion cycle*. However, this idea is complicated by the fact that the controlling factors of landscape development are rarely constant over the many millions of years it takes to go from early- to late-stage streams. A change in climate can increase runoff, thus

Dendritic drainage pattern.
The underlying rock or sediment is uniformly resistant to erosion. Common where sedimentary layers are horizontal.

Radial drainage pattern.
Streams radiate outward from a high area such as a volcano or dome.

Rectangular drainage pattern.
Streams follow fractures or faults in underlying rocks. Right-angle intersections are common.

Trellis drainage pattern.
Most streams follow valleys that parallel inclined layers of rock.

FIGURE 8.8
Four common drainage patterns.

causing a middle-stage stream to cut into its meander belt sediments and leave river terraces. Regional uplift of a stream or a drop in base level would also cause erosion by downcutting. The uplift of headwater tributaries or an increase in glacial activity could increase the amount of sediment coming into the middle stages of a river system and thus fill the valley with meander belt sediments. A rise in sea level or lowering of the land surface could also increase deposition of meander belt sediments. Additionally, a rim of resistant rock can temporarily create a secondary base level at a relatively high elevation within a river system. This causes meanders to develop upstream. Once the resistant rocks are breached by erosion, the river will cut into its old meander belt sediments and create terraces. Overall, many river systems reflect a complex mixture of present and past conditions.

WEST

EAST

Slowly eroding
tributaries of the East River

Rapidly eroding
tributaries of the
West River

Drainage divide

A.

Drainage divide Headward erosion

B.

Captured
tributaries

New drainage divide
between East and
West rivers

Original divide

Discharge of
beheaded
stream may drop
substantially.

Point of capture

C.

FIGURE 8.9

Headward erosion and stream piracy. A. Erosion is faster on the west side of the drainage divide than on the east. B. This enables several stream tributaries to cut into the divide and lengthen their drainages in a process called headward erosion. C. Stream piracy occurs when headward erosion cuts into a tributary from another stream system, thereby capturing its water flow.

Hands-On Applications

Geologists have approached the study of streams and associated landscapes in several ways. Early approaches were descriptive systems of classification of natural phenomena, which assumed cause-and-effect relations among various features. Working models were constructed to help geologists understand rivers and river systems. Eventually, enough numerical data became available from both natural and model systems to develop mathematical models, and quantitative predictions could be made. The problems in this lab illustrate both qualitative and quantitative approaches.

Objectives

If you are assigned all the problems, you should be able to:

1. Identify the following features on a map or aerial photograph: floodplain, meander, meander belt, backswamp, oxbow lake, cut bank, point bar, natural levee, cutoff.

2. Determine whether the characteristics of a stream or landscape are those of early, middle, or late stage.

3. Identify dendritic, radial, trellis, and rectangular drainage patterns, and use them to predict the general structure of the underlying rocks.

4. Use data from a gaging station to calculate the *percent probability of recurrence* or the *recurrence interval* of a particular discharge or river stage and to construct a flood-frequency curve.

5. Outline a drainage basin on a topographic map, estimate its area, and calculate its annual runoff from precipitation, evaporation, transpiration, and infiltration data.

6. Draw both cross and longitudinal profiles of a stream valley.

Problems

1. Figures 8.10, 8.11, and 8.12 illustrate segments of three major U.S. rivers that show different characteristics.

   a. Examine the maps and complete Table 8.1. Note that the maps are at different scales. The purpose of this problem is to recognize common stream features on a map and to learn which stream features are typically associated with one another. When calculating gradients for Table 8.1, measure the distances along the stream's actual course, as demonstrated in Figure 8.4.

   b. Levees are often too low to show up on contour maps. On which river(s) would you expect to find natural levees?

   c. Levees are marked on the Mississippi River map (see Fig. 6.10 for symbol). Do you think these are natural or made by people? What is your evidence?

   d. Because a river normally occupies the lowest points within its valley, small streams generally flow in a fairly direct path to join the main river. Carefully examine the somewhat complicated drainages formed by the numerous bayous (these are both permanent and intermittent streams) west of the Mississippi. Do they flow toward or away from the main river channel? Why do they flow this way?

e. Levees can be most obvious during a flood. Examine the satellite image of the confluence of the Missouri and Mississippi rivers just north of St. Louis, Missouri (Part IV opening image, p. 145). Greater St. Louis covers the whole area south of the rivers and west of the clouds. The Mississippi comes in from the northwest corner and the Missouri comes in just south of the Mississippi. Which river shows beautiful examples of levees?

Where is there a good example of a cutoff?

Where is there an artificial canal roughly half the width of the nearby river?

The curved landforms east of the clouds reflect current oxbow lakes, curved ridges from old point bars, and filled-in oxbow lakes. These are typical of meandering river landscapes.

Table 8.1
Characteristics of Three Stream Valleys

Stream Characteristics	Missouri River	Arkansas River	Mississippi River
Are there meanders?			
backswamps?			
oxbow lakes?			
cutoffs?			
River gradient within map area (in feet per mile)	(See footnote 1)		(See footnote 2)
Approximate width of floodplain (from one side to the other)			
Sketch a valley profile. Mark channel position in profile and show approximate position of floodplain.			
Importance of downcutting (high, medium, low)			
Importance of lateral erosion (high, medium, low)			
Development stage (early, middle, late)			

[1]A supplementary contour with an elevation of 750 feet crosses the river 1.5 miles below the bottom edge of the map. Use this information, together with what can be observed, to calculate the gradient.

[2]The 100-foot contour crosses the river 74 miles above the top edge of the map. Use this, along with what is observed on the map, to calculate the gradient.

Scale 1:62,500
C.I. 20 feet

FIGURE 8.10
Portion of topographic map of Leavenworth, Kansas, 15-minute quadrangle, for Problem 1.

FIGURE 8.11
Portion of topographic map of Royal Gorge, Colorado, 7½-minute quadrangle, for Problem 1.

Scale 1:24,000
C.I. 40 feet

FIGURE 8.12

Portion of Jackson, Mississippi/Louisiana, 1° × 2° quadrangle, for Problem 1.

Scale 1:250,000
C.I. 50 feet

2. Refer to Figure 8.13, a portion of the Kaaterskill, New York, quadrangle. This area, in the northeast corner of the Catskill Mountains of southeastern New York, shows a contrast in topography between the east and west sides of the map. The western part of the area is underlain by sedimentary rocks with nearly horizontal layers, and the eastern part is underlain by sedimentary rocks with steeply inclined layers.

   a. How do the drainage patterns in the eastern and western parts of the area differ?

   b. Are rocks in the eastern or western part of the area more resistant to erosion? How do you know?

   c. Which of the two creeks in the central part of the map, Schoharie Creek or Plattekill Creek (between its head and West Saugerties), has the steepest gradient?

      Which do you think is eroding headward most actively, and why?

      What is the likely future for the headwater tributaries of Schoharie Creek?

   d. Kaaterskill Creek, in the northern part of the map, has committed stream piracy and "beheaded" Gooseberry Creek. Use a colored pencil to indicate on the map the probable course of Gooseberry Creek before this tyrannous event.

FIGURE 8.13
Portion of topographic map of Kaaterskill, New York, quadrangle, for Problem 2.

Scale 1:62,500
C.I. 20 feet

0   1/2   1   2 Miles

0   1   2   3 Kilometers

3. Figure 8.14 is a map showing the Minnesota River at Jordan, Minnesota. The *gaging station* at the bridge over the Minnesota River north of Jordan is used to monitor the *stage* or level of the river as well as discharge and water quality. The stage here is the number of feet above the gaging-station elevation of 690'. Table 8.2 shows the maximum stage reached during selected years from 1935 to 2003. A relative magnitude, *m,* has been assigned to each stage and is recorded in column 3 of Table 8.2. The highest maximum stage was reached in 1965, so its magnitude, *m,* is 1; the lowest maximum stage was attained in 1935 (not shown in the table) and its magnitude is 69 (the table summarizes 69 years of records).

   a. Construct a **flood-frequency curve** as follows:

   (1) Calculate the percent probability of recurrence, P, for each of the events listed in Table 8.2, and record it in column 4 of that table. P is the percent probability that a stage of a particular height will recur—be reached or exceeded—in a given year. It is determined by:

   $$P = 100 \times m/(n + 1)$$

   where *n* is the number of years for which records were kept, and *m* is the relative magnitude of the event. For example, for a stage that ranked 30th in magnitude for a period of 69 years, $P = 100 \times 30/(69 + 1) = 42.9\%$.

   (2) Plot the data in Table 8.2 on the graph paper in Figure 8.15. Be sure to use the scale on the top of the graph, labeled *Percent probability of recurrence (P),* when plotting your numbers. Then draw a best-fit straight line through the data points to complete the flood-frequency curve. The graph paper is called probability paper because it can be used to determine the probability, or likelihood, that the event plotted on it will occur. In the example above, there is a 42.9% probability that a stage equal to that reached during the *m* = 30 event will be reached in any given year.

   b. The **recurrence interval,** or average *time* interval between similar-sized floods, is another way of describing recurrence. The recurrence interval, *RI,* is given by:

   $$RI = (n + 1)/m$$

   and *RI* = 100/*P*. Values for the recurrence interval are shown on the lower part of the graph in Figure 8.15. Note that a 42.9% probability of recurrence also means that a stage of that value would be reached, on average, about every 2.4 years. The probability of a 50-year flood (one with a recurrence interval of 50 years) in any given year is 0.02, or 2%. Thus, there is always a probability of a 50-year flood, even if one had occurred just the previous year. Conversely, there is no guarantee that a 50-year flood will occur every 50 years.

   (1) Major flooding occurred in the spring of 2001, when the maximum stage reached 33.11 feet. Use your flood-frequency curve and best-fit straight line to determine what the recurrence interval was for this flood at this location.

   c. Use the flood-frequency curve you constructed (Fig. 8.15) to determine the maximum stages for a 100-year flood (one with a recurrence interval of 100 years). From the stage values, calculate the elevation of this flood (by adding stage to 690') and, using a colored pencil, outline the 100-year floodplain on the topographic map. Note that the obvious broad valley is not the current floodplain, but was carved when much more water was coming down the river than at present. This took place about 10,000 years ago during the melting of continental glaciers and the draining of a vast lake (Glacial Lake Agassiz—see Fig. 10.6).

   The floodplain you have outlined (or similar ones based on different recurrence intervals) can be used by government agencies to plan the use of floodplains or by insurance companies to set their flood insurance rates. For example, two-year floodplains should not be used for housing or other permanent types of development and flood-insurance rates would be prohibitively expensive. People building in areas of the floodplain with longer recurrence intervals should know they can expect to be flooded sometime and should be prepared to take that risk.

Scale 1:24,000
C.I. 10 feet

FIGURE 8.14
Portion of Jordan West, Minnesota, 7½-minute quadrangle, for Problem 3.

Table 8.2
Selected Data from the Gaging Station on the Minnesota River Near Jordan, Minnesota, 1935–2003

Year	Maximum Stage, feet	Magnitude, M	P
1936	22.43	30	
1942	14.36	57	
1946	18.22	48	
1952	28.31	6	
1958	12.30	63	
1959	8.37	67	
1964	16.15	54	
1965	34.37	1	
1969	32.85	4	
1975	23.77	23	
1976	11.19	65	
1984	27.54	8	
1985	25.05	17	
1986	26.30	9	
1990	20.23	38	
1993	33.52	2	
2003	19.10	44	

Source: USGS

4. **Watershed Data on the Web.** You can investigate local flood risks, or see whether your favorite river is likely to have too little (or too much!) water for a planned canoe trip, by going to *http://water.usgs.gov/* (or link to it through *www.mhhe.com/jones6e*—see Preface). There is a lot to explore on this website, but for now click on *NWISWeb Water Data* (NWIS stands for National Water Information System) and then click on *Real-time* to see a national map of stream gaging stations. Although this exercise focuses on a particular gaging station, the steps below work for any station and your instructor may select a local stream for you to work on. Your station may lack data for certain questions.

   a. At the top, under "Geographic Area," select *Pennsylvania*. Click on *Statewide Streamflow Table,* and under "Group table by" change the tab from *Major River Basin* to *County* and hit *Go.* Under Allegheny County, find (don't click!) "Monongahela River at Elizabeth, PA." What is the stream flow (discharge) in cubic feet per second? How does this compare to the long-term median flow for this date?

   b. Click on the link for this station, *03075070.* Is the flow to this station entirely natural or is it affected by such factors as hydroelectric plant discharges or other dam and lock controls?

   c. Scroll down to the graph of discharge. From the graph, what are highest and lowest discharges over the past week? How does the weekly flow compare overall to the median discharges plotted on the graph?

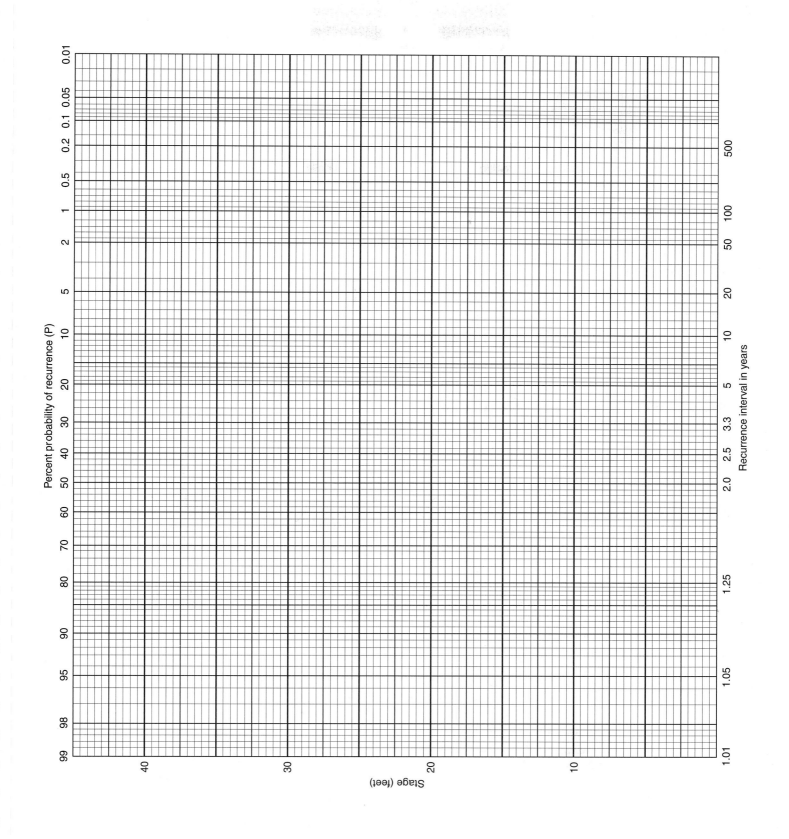

FIGURE 8.15
Graph paper for Problem 3.

d. Back at the top, change "Available data for this site" from *Real-time* to *Station site map*. Describe the location of the gaging station relative to the nearest major city (use the *Zoom In* or *Zoom Out* functions under the maps).

e. At the top, change "Available data for this site" from *Station site map* to *Surface Water: Daily Streamflow*. Under "Retrieve data from," put "1995-01-01 to today's date" (use the proper data format). Ensure the *Graphs of data* button is selected. Hit *submit*. In which year was the discharge the highest? The lowest?

The tick marks on the time axis should divide the year into quarters, each three months long. Does the discharge vary consistently by season? If so, when are the highest discharges?

f. To construct a flood-frequency curve and calculate recurrence intervals, you need to find the maximum flood for each year back through time. Under "Available data for this site" go to *Surface water: Peak streamflow*. Under "Output formats" click on *"Table."* This gives the required numbers. How many years make up the record for this station?

Compared to the 67-year record used in Problem 3, would you expect the data from this station to do a better or worse job of constraining the magnitude of a 100-year flood event? Why?

Unless directed to do more by your instructor, finish this problem off by listing the year in which the peak discharge was the highest and the year in which it was the lowest.

To find out about the water quality of this river, change the "Available data for this site" to *Water-Quality: Discrete samples*. Under "Retrieve Water quality samples for Selected Sites," check *Inventory of water-quality data* and, after changing the selection box from *For printing* to *With retrieval*, hit *Submit*. You will see a table listing the various aspects of water quality available for this site. By checking one or more boxes to the left of the table and clicking on *Submit*, you can see the actual data.

In Greater Depth

5. Figure 8.16 is a portion of the Urne, Wisconsin, quadrangle to be used for the following problems.

   a. On a separate piece of paper, draw a longitudinal profile from the mouth of Deer Creek to its head at the pond near the center of Section 3 (labeled with a red 3). Find the length of the stream (or the profile) by following all the curves and measuring the actual distance along the channel; mark and label those points where a contour crosses the stream. This can be done most easily by using the edge of a piece of paper. Use a vertical scale of 1 inch to 100 feet for the profile.

   What is the vertical exaggeration? Show your calculations.

   In what part of the stream is the gradient greatest? How do you know?

FIGURE 8.16
Portion of Urne, Wisconsin, 7½-minute quadrangle, for Problem 5.

Scale 1:24,000
C.I. 20 feet

0   ½   1 Mile

0   .25   .5   1 Kilometer

What is the shape of the profile? Is it concave or convex upward? Is it smooth, or are there abrupt changes in elevation such as might be caused by a waterfall?

Does the shape of the profile suggest that the stream has or has not attained a near balance between erosion and deposition, and how did you tell?

b. Outline the stream channels and drainage basin of Deer Creek and its unnamed tributaries. The boundary of a drainage basin is a drainage divide, the line of separation between adjacent drainage basins. It can be a ridge, summit, or tract of high ground. Water falling within the drainage basin will flow toward the mouth of the stream; water falling outside the drainage will flow into another stream system.

c. Estimate the area of the drainage basin in square miles by adding the number of whole sections and fractions of sections that fall between the divides (one section = 1 square mile).

Using your estimate, calculate the area of the drainage basin in square feet (1 mile = 5280 feet). Show your work.

d. The average annual precipitation for this area is 29.1 inches (2.4 feet). Calculate the average volume of water that falls on the drainage basin per year in cubic feet (volume = area × annual precipitation). Show your work.

e. The average volume of water that runs off and flows to the mouth of Deer Creek per year is the average annual discharge of Deer Creek. Calculate the average annual discharge, assuming the following different conditions:

(1) Infiltration loss, evaporation, and transpiration are all zero, and no groundwater is contributed from outside the drainage basin.

(2) Infiltration loss = 2 inches per year, evaporation + transpiration = 15 inches per year, and no groundwater comes from outside the drainage basin. Show your calculations.

The Urne quadrangle is in the unglaciated or driftless area of Wisconsin. Note that little flat or gently sloping land can be found in the Urne quadrangle, except in the floodplain of the Mississippi River. This is one indication that the area has not been glaciated and that streams have been the principal eroding agents. If glaciers had covered the area, the topography would be more subdued: hilltops would have been eroded, valleys would have been wholly or partly filled with glacial drift, and surface drainage would not be as well developed.

CHAPTER 9

Groundwater and Groundwater-Influenced Landscapes

Materials Needed

- Pencil and eraser
- Calculator
- Ruler

Introduction

Water beneath the Earth's surface that occupies open spaces in rock or sediment is called **groundwater.** If the spaces are sufficiently connected, it can flow from one place to another. This enables us to pump water from a well without it immediately running dry, but it also means that we can contaminate important freshwater resources through careless waste disposal. In carbonate rocks, naturally acidic water flowing between grains and in cracks dissolves the rock and enlarges the open spaces. These openings can grow in size from a few millimeters up to huge cave systems.

Groundwater infiltrates from the surface and is part of the water cycle, as illustrated in Chapter 8. Typically, surface water percolates downward through unconsolidated soil and sediment until it enters a zone in which all of the interconnected spaces between sediment grains are filled with water, as illustrated in Figure 9.1. Near the surface, in the **unsaturated zone,** or **zone of aeration,** these **pore spaces** are mostly filled with air. Below the **water table,** interconnected pore spaces are filled with water in the **zone of saturation.** In humid regions, the elevation of the water table generally rises below hills and falls below valleys; its shape is commonly a subdued replica of the land surface.

The problems for this chapter illustrate how the shape of a water table can be determined in areas where lakes are abundant, how the flow direction of groundwater can be determined and used to predict the movement of pollutants, how water-bearing zones differ, and how groundwater can shape the surface topography of areas underlain by carbonate rocks.

Occurrence and Movement of Groundwater

Porosity

The **porosity,** or percentage of open (pore) space in the rock or sediment, determines the volume of water in the zone of saturation. In most rocks or sediments, the only pore spaces are those between adjoining particles or the narrow openings along fractures (cracks). In carbonate rocks and some lava flows, however, there may be large open spaces, and water may actually flow as underground rivers.

Permeability and Flow

The rate (velocity) at which groundwater flows is determined by **permeability** (ease of conducting water), **effective porosity** (the percent of open space through which the groundwater actually flows), and **hydraulic gradient** (for this lab, the slope down which the groundwater flows). Highly permeable material includes rock or sediment with large, interconnected pore spaces; low permeability materials have poorly interconnected pores. Groundwater flows fastest moving down a steep slope through a highly permeable material like sand or gravel. It moves much more slowly through an impermeable material like shale, especially when there is a low hydraulic gradient. The rate at which water flows through all such porous materials is expressed by the following equation derived from Darcy's law:

$$v = \left(\frac{K}{P}\right)\left(\frac{h}{l}\right) \qquad \text{Eq. 9.1}$$

where v is the velocity of groundwater flow (also known as the *seepage velocity*); K is the **hydraulic conductivity,** which depends on the nature of the medium and is directly related to permeability; P is the effective porosity; h is the **head,** which in this

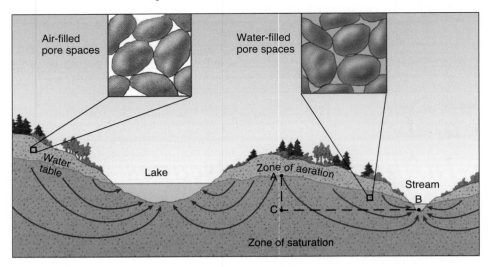

FIGURE 9.1

Cross section of an unconfined aquifer illustrating the zones of aeration and saturation, the water table, and groundwater flow paths. The exploded views show that air fills the pores in the zone of aeration while water fills the pores in the zone of saturation. Water falling on the land surface flows straight down through the zone of aeration. Once in the zone of saturation, it flows toward nearby rivers and lakes. Groundwater builds up to higher elevations away from where it exits to rivers, lakes, and springs, thus building a hydraulic gradient to drive groundwater flow. This combination of sideways flow plus more water coming in from above (from rain) creates the curved paths. Water flowing from A to B has a hydraulic gradient equal to the head (A to C) divided by the horizontal flow distance (C to B).

Example

If, in Figure 9.1, the head *(A-C)* is 75 m, the horizontal distance traveled *(C-B)* is 1.2 km, the hydraulic conductivity of the sediment through which the water moved is 3.0×10^{-3} cm/sec, and the effective porosity is 20%, calculate the velocity of the water discharging into the stream in cm/sec.

Because the answer should be in cm/sec, first convert *A-C* and *C-B* to centimeters: 75 m = 75 m × 100 cm/m = 7500 cm = 7.5×10^{3} cm; 1.2 km = 1.2 km × 1000 m/km × 100 cm/m = 120,000 cm = 1.2×10^{5} cm. Then use Equation 9.1, where *A-C = h, C-B = l,* the hydraulic conductivity is *K,* effective porosity is *P,* and the velocity is *v:*

$$v = (K/P)(h/l)$$
$$v = (3 \times 10^{-3} \text{ cm/sec}/0.2)(7.5 \times 10^{3} \text{ cm}/1.2 \times 10^{5} \text{ cm})$$
$$v = 9.4 \times 10^{-4} \text{ cm/sec}$$

What would the velocity be in centimeters per day? First determine the number of seconds in a day. One day has 24 hours, each hour has 60 minutes, and each minute has 60 seconds. Therefore, there are 24 hours/day × 60 minutes/hour × 60 sec/min = 86,400 sec/day = 8.64×10^{4} sec/day. The velocity per day is, $v = 9.4 \times 10^{-4}$ cm/sec × 8.64×10^{4} sec/day = 81 cm/day.

lab is taken as the difference in elevation between the top and bottom of the flow path; and *l* is the horizontal distance over which the groundwater flows. The hydraulic gradient is *h/l.* Figure 9.1 shows a common situation in a humid area in which water flows underground, following a curved path, and discharges into a stream or lake. Water going from point A to point B, Figure 9.1, drops a distance *A-C* and travels a horizontal distance *C-B: A-C* is the head, *h,* and *C-B* is the length, *l,* in Equation 9.1.

Aquifers

An **aquifer** is a porous and permeable sediment or rock from which a useful amount of groundwater can be obtained. Sand and gravel, sandstone, limestone, dolomite, basalt flows, and some *fractured* granites and metamorphic rocks are good aquifers.

An **unconfined aquifer,** like the one in Figure 9.1, is one in which movement of water is unrestricted, and the aquifer is recharged by water percolating downward from the overlying surface. Sand and gravel make excellent unconfined aquifers.

A **confined aquifer** is overlain by, or sandwiched between, comparatively impermeable **confining layers,** as shown in Figure 9.2. Common confining layers are clay and shale. Confined aquifers are recharged where they are exposed at the surface *(the recharge area),* and by whatever water can leak through the confining beds.

Springs

Springs occur where groundwater comes to the surface. This may happen where fractures, faults (fractures on which movement occurs), caves, or a contact between a permeable layer and underlying impermeable layer intersect the surface.

Human Use

We make use of groundwater by drilling wells and pumping it to the surface. We also contaminate groundwater by carelessly disposing of contaminants in a way that allows them to enter the groundwater system.

Wells

A well is a hole dug or drilled into the ground for the purpose of obtaining groundwater (Fig. 9.3). Wells may penetrate unconfined or confined aquifers. In either case,

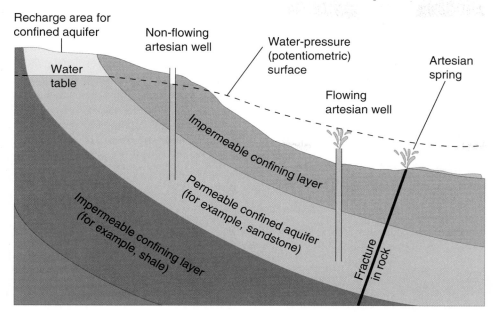

FIGURE 9.2
Cross section illustrating a confined aquifer system with recharge area, water-pressure surface, and flowing and non-flowing artesian wells.

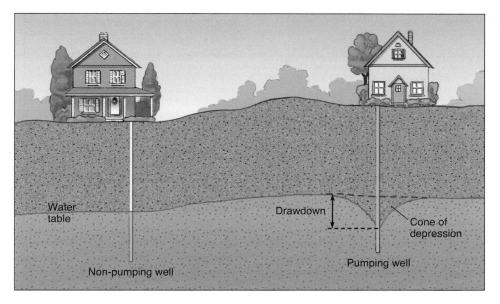

FIGURE 9.3
Cross section illustrating a non-pumping and pumping well in a simple unconfined aquifer. Pumping causes the water table to be drawn downward, forming a cone of depression.

water from the aquifer flows toward the well, an area of low relative pressure. If the aquifer is unconfined, the hole will fill with water up to the level of the water table (Fig. 9.3).

Confined aquifers have their recharge areas at higher elevations than most of the aquifer (Fig. 9.2). This pressurizes the confined water in the aquifer. If you drill a well into a confined aquifer, water generally rises above the level of the aquifer, forming an **artesian well** (Fig. 9.2). The level to which water in a well rises above the confining layer defines the **potentiometric** or **water-pressure surface** (Fig. 9.2). An **artesian spring** forms where water moves up along a fracture (Fig. 9.2).

Water is extracted from most wells by pumping. Pumping causes the level of the water and the pressure in the well to drop, allowing more water to flow into the well. If the aquifer is unconfined, continuous pumping causes a **cone of depression** to develop on the water table around the well, as shown in Figure 9.3. This also occurs when you use a straw to quickly suck the flavoring from a crushed-ice slurpee.

If the amount of water withdrawn from an unconfined aquifer exceeds the amount added by recharge, the water table throughout the region will drop. Excessive withdrawal from confined aquifers will lower the potentiometric surface.

In extreme cases, extraction of water from pore spaces in the aquifer may cause the pore spaces to collapse. As a result, the aquifer may be permanently damaged. In some cases, porosity in the aquifer or in confining beds decreases so much that the land surface subsides. For example, so much water has been withdrawn from the lake sediment on which Mexico City is built that the opera house has settled 3 m, and parts of the city have subsided as much as 8 m.

Fluids with Different Densities

Lighter, less dense fluids tend to float on heavier, more dense fluids, whether in a mixing bowl in the kitchen or in a permeable rock below the surface. Four natural fluids, listed in order of increasing density, are natural gas, oil, fresh water, and salt water—if all four are present, gas will be on the top and salt water on the bottom.

Salt water fills the pore spaces in sediment below the oceans, and some continental aquifers contain saline water. In coastal areas, saline and fresh groundwater come in contact, as shown in Figure 9.4. Because salt water is more dense, it underlies fresh water in the groundwater system. Active removal of fresh water by pumping will change the position of the salt water interface and may cause infiltration of salt water into some wells, a problem known as salt-water encroachment.

Oil and natural gas, being insoluble in water and lighter than either saline or fresh groundwater, rise through permeable rocks until they reach an impermeable barrier. If that barrier cannot be passed, the gas and/or oil is trapped, and an oil or gas pool will form. As shown in a simple example in Figure 9.5, when water, oil, and gas are all

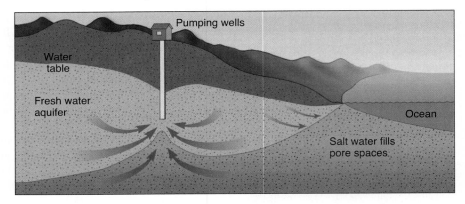

FIGURE 9.4
Cross section illustrating saltwater encroachment caused by excessive withdrawal of fresh water.

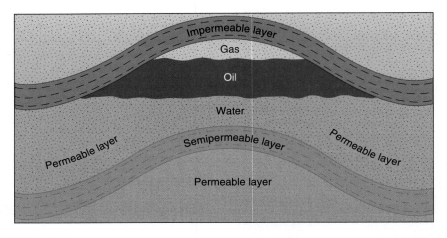

FIGURE 9.5
Oil and gas rise in permeable rocks until they encounter an impermeable layer and are trapped, as illustrated in this cross section.

present, gas, being the lightest, rises to the top, while water sinks to the bottom and oil settles in the middle.

Contamination

Contamination of groundwater can be a severe problem because it is generally not detected until it affects a well or enters a stream. Once contaminated, an aquifer is difficult and expensive to purify. Figure 9.6 illustrates some of the ways in which groundwater can be contaminated by human activities. Many contaminants are somewhat soluble in water, allowing them to disperse and making it difficult or impossible to recover them. In some cases, it is possible to dispose of such materials by injecting them into deep, permeable rocks containing dense salt water that is isolated from freshwater aquifers. Insoluble contaminants may segregate within the system on the basis of density. For example, waste oil disposed of at the surface would percolate down to the top of the water table, but would not mix with underlying fresh water. If the oil does not migrate very far laterally, it may be possible to reclaim it.

Groundwater in Carbonate Rocks

Although most of the world's groundwater occurs in porous rock, it can also be present in fractures or even underground streams. This is particularly true of groundwater in carbonate rocks because carbonate minerals—especially calcite—dissolve in naturally acidic groundwater. Groundwater is slightly acidic because it contains dissolved carbon dioxide derived from the atmosphere or organic acids released by the decay of organic matter in soil. Water combined with dissolved carbon dioxide makes carbonic acid (H_2CO_3), which then acts on carbonate minerals. The overall reaction is:

$$H_2O + CO_2 + CaCO_3 = Ca^{+2} + 2HCO_3^{-1}$$

As a result, calcium (Ca^{+2}) and bicarbonate (HCO_3^{-1}) ions are carried away in groundwater, and cavities form in carbonate rocks.

Caves, Disappearing Streams, and Sinkholes

The largest subsurface cavities are **caves,** which commonly develop as groundwater

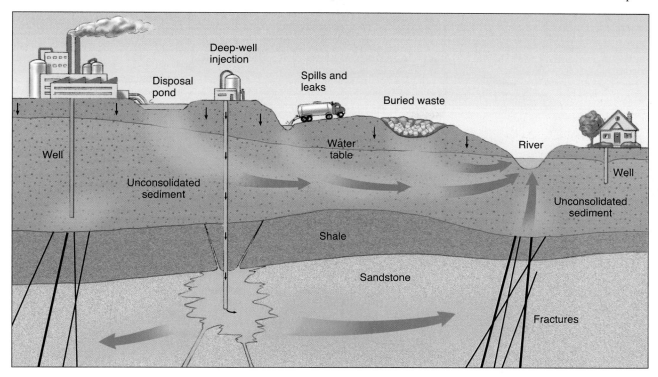

FIGURE 9.6
This cross section illustrates a number of ways in which groundwater can be contaminated by human activity.
Source: USGS Circ. 875.

flows along sedimentary layers or fractures. Where dissolution occurs along horizontal surfaces, such as sedimentary layers, caves have dendritic or braided patterns when viewed on a map. Where fractures control water flow, a more regular pattern is common, with chambers intersecting at consistent angles. Most caves form at or below the water table, but they can also form above it. Caves serve as channels for ground-water flow and act as underground rivers. **Disappearing streams** are surface streams that plunge underground to become underground rivers. **Sinkholes** are depressions formed by the dissolution of carbonate rock beneath the overlying soil or by the collapse of portions of near-surface caves.

Karst Topography

Areas underlain by carbonate rocks in which dissolution has been active develop a distinctive surface appearance known as **karst topography.** Figure 9.7 shows features of karst topography in various stages of development. In the early stages, scattered sinkholes appear, and the surface drainage is disrupted as some streams disappear underground. Continued cave collapse leads to the presence of valleys, which form as sinkholes enlarge and join. Eventually, all that may remain are haystack-shaped hills consisting of rock that has not dissolved. It has been estimated that 20 percent of the Earth's land surface and 40 percent of the U.S. east of Oklahoma show some manifestation of karst topography. In particular, Kentucky, southern Indiana, and much of Florida are famous for their sinkholes and karst topography.

FIGURE 9.7

The evolution of karst topography. A. Scattered sinkholes and disappearing streams characterize the early stage. B. Continued dissolution and collapse of caves leads to the formation of valleys. C. Eventually, all that remains are scattered limestone hills.

Hands-On Applications

Because groundwater and the rock or sediment through which it moves are not directly visible, our ability to understand and predict its behavior is strongly dependent on models or hypotheses. These models can be tested by various means. Where carbonate rocks have been dissolved by groundwater, direct observations can be made in caves and at large springs. In most areas, however, data are obtained from wells, which are samples of a very small volume. In this lab, you will learn some of the methods used by hydrogeologists to interpret and predict the movement of groundwater, and its relation to the rock or sediment that contains or confines it.

Objectives

If you are assigned all the problems, you should be able to:

1. Use water-table contours and flow lines to predict groundwater flow.

2. Use water-level elevations, whether in lakes or wells, to contour the water table or draw it on a cross section.

3. Determine a hydraulic gradient.

4. Recognize distinctive karst topography and the following features on a topographic map: sinkhole, disappearing stream, solution valley.

5. Explain how water wells in confined and unconfined aquifers differ.

6. Use water-table and water-pressure-surface information to determine how deep a well must be drilled to encounter water from confined and unconfined aquifers.

7. Predict how fluids of different densities would behave in a groundwater system.

Problems

1. Figure 9.8 is part of the Lakeside, Nebraska, quadrangle, and is part of the Sand Hills area of western Nebraska, an area composed of old sand dunes that now are mostly stabilized by vegetation. The dunes themselves have irregular surfaces with many small depressions. This makes for a complex, "busy" contour pattern. Some of the low areas between the dunes contain small lakes. Because the sand is so permeable, these lakes form where the water table intersects the surface, and their elevation, printed on most lakes, is that of the water table.

   a. Draw a contour map of the water table using the lake elevations in the same way you used spot elevations to make a topographic map in Chapter 6, Problem 4. Use a contour interval of 5 feet and start with the east side of the map by drawing the 3900-foot contour.

   b. In what direction does the water table slope, and what is the average gradient in feet per mile?

   c. Assuming the gradient just determined is the hydraulic gradient for groundwater entering the system at Ellsworth and discharging near Lakeside, calculate the velocity, in cm/sec, at which it would flow between these points. Assume a *hydraulic conductivity, K,* of 1×10^{-3} cm/sec and an *effective porosity, P,* of 20%. Hint: Use Equation 9.1, but first convert your hydraulic gradient from units of feet/mile to feet/foot; once this is done, the hydraulic gradient will be a unitless number.

d. If the distance between Ellsworth and Lakeside is 7 miles (11.3 km = 1.13×10^5 cm), how long would it take for the groundwater to make this journey?

e. Note the windmill in the SW1/4, Sec.22, T24N, R44W (southeast of Lakeside) and its map symbol; the surface elevation is also given. Many other windmills are shown on the map by symbol only. The windmills are connected to water wells and used to pump groundwater to fill water tanks for cattle.

Locate the windmill in the NW1/4, Sec.19, T24N, R43W. How high must the water be raised by the pump to fill the stock tank? Explain your answer.

2. Mammoth Cave, in Kentucky, is one of the best-known caves in the United States—and the largest, with more than 560 km (348 miles) of passageways. The entrance to the cave is near the northeast corner of the topographic map in Figure 9.9, where red lines mark roads. Three areas with different topography are evident in this map. In the northeast, a series of ridges, some with nearly flat tops, is dissected by valleys without streams. In the northwest, the valleys are not as deep, but they do contain streams. Both areas extend south to a sharp topographic break, known as (but not labeled on the map) the Dripping Springs Escarpment, which is just north of the Louisville and Nashville Road shown on the map. The third area, south of the escarpment, is characterized by numerous depressions.

The area is underlain by nearly horizontal layers of sedimentary rock of two kinds: (1) sandstone (with some interbedded shale), which is essentially insoluble in groundwater; and (2) limestone, which is soluble in groundwater. The sandstone is on top and at higher elevations than the limestone.

a. Carefully study the distribution of depression contours on the map (a hand lens or magnifying glass may help). Make a simple outline around the large area in which depression contours are generally *not* found. Overall, are depression contours more likely to be found in areas with elevations higher than about 700 feet or lower? In which of the two types of rock in the area, sandstone or limestone, are depressions more likely to occur?

b. Why are there no streams in the northeast part of the map?

c. What evidence suggests that there once were streams in the northeast? Hint: Note the pattern of the valleys.

Scale 1:62,500
C.I. 20 feet

0   ½   1   2 Miles

0   1   2   3 Kilometers

FIGURE 9.8
(Page 177) Portion of Lakeside, Nebraska, 15-minute quadrangle, for Problem 1.

   c. Horn.

   d. U-shaped valley.

   e. Arête.

   f. Hanging valley.

   g. Tarn.

   h. Paternoster lakes.

2. Refer to the aerial photographs of North Crillon Glacier (Fig. 10.8) and to the map of part of Glacier Bay National Park, southeastern Alaska (Fig. 10.9). North Crillon Glacier also appears on the map.

   a. Look at Figure 10.8 with a stereoscope. North is marked.

   (1)   Look at the small white glacier north of North Crillon Glacier. What geological term describes the valley occupied by the small glacier compared to the larger valley hosting the larger glacier?

         What evidence suggests that the small glacier was once thicker? While you are looking, note that the small glacier forms an *ice fall* near the top of the photo.

   (2)   Examine the valley wall north of North Crillon Glacier. Draw a line indicating the height to which the glacier ice may have once extended. Did the North Crillon Glacier fill its valley up to the level of the small glacier, or is it more likely that the small glacier made another ice fall as it entered the larger valley?

   (3)   Examine the terminus (end) of the North Crillon Glacier. Why is a big piece missing from its north end?

         Although you cannot see a meltwater stream, there is strong evidence that summer meltwater transports a great deal of sediment along the southwestern edge of the glacier. What evidence can you see?

   (4)   What are the dark streaks on the upper end of North Crillon Glacier?

         Approximately how many are there?

         Assuming the above number is correct, infer how many tributary glaciers flow into North Crillon Glacier.

   (5)   Describe why the glacier is darkest near its terminus.

FIGURE 10.8
Stereophotographs of North Crillon Glacier, Glacier Bay National Park (see Fig. 10.9 for location). Scale is 1:40,000.

(6)   Note the cracks, or *crevasses*, near the terminus. What is their orientation relative to the direction of ice flow?

Did they form from tension (pulling apart) or compression (pushing together)? Explain your answer.

b. As you can see in the map (Fig. 10.9), North and South Crillon glaciers flow southwest, bend 90° as they enter a northwest-trending valley, and then go either northwest or southeast. Incidentally, the northwest-trending valley marks the position of one of the major geologic faults of southeastern Alaska. Earthquakes along the Fairweather Fault occasionally jar the area.

(1)   Starting with the ridge between North and South Crillon glaciers, trace the drainage divide that defines the drainage basin of the North Crillon Glacier (Chapter 8 explains drainage basins). Look for the sharp-edged arêtes and horns to help guide your way; parts of the divide are snow-covered.

Approximately how many tributary glaciers, large and small, can you spot flowing into the North Crillon Glacier? Ignore the smallest "buds." Your answer probably will not match what you got from the air photos because closely spaced medial moraines tend to merge as additional glaciers squeeze into the flow.

(2)   Cite evidence to support or refute the suggestion that glaciers once flowed down the creeks north of Lituya Bay (i.e., Fourmile, Eagle, and Portage creeks).

(3)   What evidence suggests that Lituya, North Crillon, and South Crillon glaciers were more extensive in the past?

Based on this evidence, mark their probable former extent on the map.

What type of glacial feature is Lituya Bay?

What type of glacial feature is at least partly responsible for the shallow water and major features at the mouth of Lituya Bay and for the ridge curving around the southern end of Crillon Lake?

c. The brown stippled areas on the glaciers in Figure 10.9 represent rocky sediment on the surface of the ice. Find and label representative lateral, medial, and end moraines.

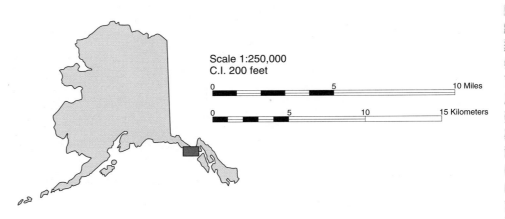

Scale 1:250,000
C.I. 200 feet

0   5   10 Miles

0   5   10   15 Kilometers

FIGURE 10.9
(Page 195) Portion of Mt. Fairweather, Alaska/Canada, 1° × 2° quadrangle, including part of Glacier Bay National Park.

3. Figure 10.10 is a portion of the Glacial Map of the United States East of the Rocky Mountains. It shows the kinds of glacial deposits present at the surface. If you were to go to Indiana, for example, and stand on one of the belts marked in green, you would know by consulting the map explanation on the left that you were standing on an end moraine of Wisconsin age. Because this map shows only surface deposits and features, it displays only the most *recent* glacial event to have affected individual areas. Two major periods of continental glaciation, both within the Pleistocene Epoch (see Figure 13.7), are well represented on this map. Most of the map shows, in shades of green, deposits of Wisconsin age, because this is the most recent glaciation. The pink areas in southern and southwestern Illinois were not affected by Wisconsin-age glaciers, but were affected by glaciers of the earlier Illinoian age. The position of the ice front shifted back and forth considerably during glacial times, but each glacial age ended with total retreat.

a. What glacial feature is represented by the dark green, curved bands, some of which are named?

Was the front of the ice advancing, retreating, or staying more or less in place as these features were being formed?

What glacial feature, shown in light green, occurs between the dark green bands in Illinois, Indiana, and Ohio?

Locate the dark green named bands in eastern Indiana. The one on the northeast is labeled *Fort Wayne,* the one on the southwest *Union City.* Which is older, the Fort Wayne or the Union City? How can you tell?

b. What do the gray areas surrounding especially lakes Huron and Erie represent?

What do the bright yellow areas represent? Hint: Examine the branching patterns seen especially in Indiana.

Today the Great Lakes drain to the east via the St. Lawrence Seaway out to the Atlantic. As the ice sheets melted, did the water always drain *only* east? What evidence supports your answers?

c. The familiar lakes, such as Lake Michigan, are shown in light blue on the map, but other lakes, such as Lake Wauponsee or Lake Watseka (southwest of Chicago), may be less familiar. What are these lakes?

Explain how and when they came to be, and what evidence for them you might find in the landscape today.

4. All of the northern Midwest except for the southwestern part of Wisconsin and a small part of adjacent Illinois was covered by continental glaciers several times during the Pleistocene Epoch (Fig. 10.10). The aerial photograph in Figure 10.11 includes the boundary between glaciated and unglaciated areas. The red area on the east is hilly and partially forested. The area on the west, with the circular irrigated fields, is a flatter terrain that is mostly cropland.

    a. Which side has been glaciated, east or west, and how do you know? Hint: Consult both Figure 10.10 and Figure 10.11.

    b. What type of glacial landform marks the former outer edge of the continental glacier?

    c. What kind of glacial sediment would you expect to find in the eastern part of the area?

    d. What type of glacial landform occurs in the western part of the area?

    e. What kinds of glacial sediment make up this type of landform?

5. Figure 10.12 represents an area of southern Michigan just south of Jackson and located in one of the yellow areas of Figure 10.10. The Jackson area consists mainly of glacial deposits associated with stagnant (nonmoving) ice. These are mostly deposits from meltwater that flowed on, in, or at the base of ice that later melted away.

    a. How did the several lakes—such as Mud, Crispell, Skiff, and so forth—probably form?

    b. What type of glacial feature is Blue Ridge, and how did it form?

    c. What type of glacial feature is Prospect Hill, and how might it have formed?

    d. Newer, 1:24,000-scale maps of the north half of this area show crossed-shovel symbols in several places. What resource do you think is being extracted?

    Circle several areas in the northern half of the map where you think comparable resources might be located.

FIGURE 10.10
Portion of glacial map of the United States east of the Rocky Mountains.

EXPLANATION
AREAL COLORS

Lacustrine sediments
Mainly sand, silt and clay, deposited in lakes dammed by glacier ice or by outwash sediments. Includes areas of till or bedrock. Labels show the names of lakes except ancient Great Lakes.

Outwash sediments
Mainly sand and gravel, deposited by proglacial streams. Includes some alluvium; in Kansas, Nebraska, South Dakota, North Dakota, and Montana, includes alluvium with little or no outwash. Arrow shows direction of flow.

Ice-contact stratified drift
Mainly sand and gravel, occurring as kames, eskers, kame terraces, pitted plains and collapsed stratified drift.

End moraines of Wisconsin age
Mostly till. Local names are lettered on or beside the moraines.

Drift, other than end moraines and outwash, of Wisconsin age
Chiefly till. Includes many areas of bedrock and older drift.

PAYSON
End moraines of Illinoian age
Mostly till. Local names are lettered on or beside moraines.

Drift, other than end moraines and outwash, of Illinoian age
Chiefly till. Includes many areas of bedrock.

Drift of Kansan age
Chiefly till. Includes large areas of bedrock.

Drift of Nebraskan age
Patches of till and scattered erratics.

Area not glaciated

Streamline features

Striation direction

Strandlines

0	25	50	100 Miles

0	50	100	150 Kilometers

FIGURE 10.11

High-altitude, color-infrared aerial photograph of the Coloma, Wisconsin, area (scale 1:58,000) for Problem 4. The circles are irrigated fields. Red indicates growing vegetation. Greens and lighter shades reflect land with little vegetation. Photo location is shown in Figure 10.10.

FIGURE 10.12

(Page 201) Portion of Jackson, Michigan, 15-minute quadrangle topographic map.

6. **Glaciers on the Web:** Go to *http://nsidc.org/glaciers* (or link to it through *www.mhhe.com/jones6e*—see Preface) and click on *General Information*, then *Q&A*.

a. Click on *What is a glacier?* What percentage of the world's land area is occupied by glaciers today?

What percentage was occupied during the last Ice Age?

b. Click on *Where are glaciers located?* What is the total area covered by glaciers around the world?

What *percent* of this is in Antarctica? Show your work.

In the United States?

In Africa?

Does Iceland have more or less ice than Greenland?

c. Click on *What types of glaciers are there?* What are the differences between an *ice sheet, ice cap,* and an *ice field?*

d. Click on *Do glaciers affect people?* What essential resource do some glaciers provide people?

Click on *Are glaciers dangerous?* What are the main hazards for people living near glaciers?

e. Click on *How do glaciers reflect climate change?* How can glaciers give us information regarding the history of climate over the past several hundred thousand years?

What might they be telling us about climate change over the past century?

Under *Why do glaciers move?* you can see a striking example of glacial retreat.

Another site of interest is the U.S. Geological Survey site *http://ak.water.usgs.gov/glaciology,* which describes the "benchmark" glacier program. This program monitors climate, glacier motion, mass balance, geometry, and stream runoff at three glaciers. Many other interesting sites can be reached through links from this site.

In Greater Depth

7. The huge Glacial Lake Agassiz (Fig. 10.6) once drained south out the Minnesota River. A portion of this river is shown in Figure 8.14 as part of Problem 3 in Chapter 8. If you worked this problem, you learned that the 100-year flood plain is located roughly at the level of the 730-foot contour.

    a. Use Figure 10.13 to create a topographic profile along the line A-A' shown in Figure 8.14.

    b. Locate and mark (on Fig. 10.13) the position of the floodplain for the *modern* Minnesota River.

    c. Do you think the Minnesota River carried the same average flow when it was draining Lake Agassiz as today? Mark reasonable limits for the ancient floodplain on the profile you drew on Figure 10.13.

A   **A'**

FIGURE 10.13

8. Figure 10.14 is a map that summarizes pre-1970 data for Grinnell Glacier, a cirque glacier in Glacier National Park (west center of Fig. 10.7).

    a. The arrows or vectors shown on the map summarize the total movement of a number of rocks exposed on the surface of the glacier, as indicated in the *Explanation*. The lengths of the vectors depend on both the time over which observations were made and the *rate* at which the ice moved. What if we wanted to know whether the rate varied from year to year, or whether it varied from place to place in the glacier? Table 10.1 contains the data from which the length of the vectors was determined, and it shows the average annual rate of movement for one of the rocks.

    (1)   First, fill in the table by calculating the average annual rate of movement for all the rocks.

    (2)   Then separate the rates into four categories ranging from slowest to fastest (for example, if the difference between the slowest and fastest rate is 40 feet/year, and the slowest rate is 25 feet/year, the categories would be 25 to 34 feet/year, 35 to 44 feet/year, 45 to 54 feet/year, and 55 to 65 feet/year). List your categories next to Table 10.1.

Table 10.1
Movement of Marked Rocks on Grinnell Glacier

Source: Data from A. Johnson, *USGS Professional Paper 1180*, 1980.

Rock Number	Period	Movement (feet)	
		Total	Annual average
47-1	1947–57	380	38
47-2	1947–53	215	
50-1	1950–64	530	
50-2	1950–69	700	
52-1	1952–62	485	
52-2	1952–56	210	
52-3	1952–59	285	
58-1	1958–69	385	
59-1	1959–64	170	
59-2	1959–68	355	
59-3	1959–69	355	
59-4	1959–65	275	
59-5	1959–66	320	
63-1	1963–66	110	
63-2	1963–69	190	
64-1	1964–69	185	
65-1	1965–69	170	
65-2	1965–69	180	
65-3	1965–69	185	
66-1	1966–68	80	

(3)   Using colored pencils or an equivalent, highlight on the map those vectors representing the highest rates in red, second highest in yellow, third in green, and slowest in blue.

(4)   Does the ice seem to move more rapidly in some parts of the glacier? If so, give a possible explanation.

b.   It is clear from Figure 10.14 that Grinnell Glacier has receded since 1887, when it was first described by the George B. Grinnell party (see sequence of "edge of glacier" lines). However, the position of the terminus of such a small glacier over a short time period may or may not reflect the true mass balance of the glacier. The terminus position can be strongly influenced by rate of ice movement and by bedrock topography. Another way to look at changes is by periodic observations of ice-surface elevations, such as those reported in Table 10.2.

The data in Table 10.2 are mean (average) elevations of five segments of the line B-B' shown on the map. The segments are designated with the distance from plane table benchmark 6425, located near the intersection of lines A-A' and B-B' on Figure 10.14. For example, the column headed 100–500' lists the mean elevations of the segment of line B-B' between 100 and 500 feet south-southwest of the point labeled 6425.

(1)   Use these data to construct a graph on Figure 10.15 that illustrates how the average elevation of the ice surface in just the 500- to 1000-ft interval has changed with time. Plot mean elevation on the vertical *y*-axis and the date (year) on the horizontal *x*-axis. Draw lines between adjacent data points to emphasize the changes. Although these data represent one small area of the glacier, changes elsewhere on the glacier appear to be similar, so you can assume your graph represents the glacier as a whole.

(2)   How has the elevation of the glacier changed during this short time interval?

FIGURE 10.14
Map of Grinnell Glacier, Glacier National Park, Montana.

(3)   Is this generally consistent with the changes in position of the terminus shown on the map?

(4)   Does this imply that Grinnell Glacier has advanced or retreated during this period?

(5)   Has the *rate* at which elevation changed been consistent throughout the period? If not, indicate when it changed, and identify those time periods when it was generally consistent. When considering rate changes, look at rates established over periods of five or more years, not year-to-year changes. (Hint: Rate is expressed graphically by the slope of the line.)

c. An interesting website showing photographs of Grinnell Glacier at various times between 1910 and 1997, and an aerial photograph showing the boundaries of the glacier in 1850, 1937, 1968, and 1993, is at *www.nrmsc.usgs.gov/ research/grinnell.htm.*

Table 10.2
Mean Ice-Surface Elevations of Segments of Profile B–B', Figure 10.13

Source: Data from A. Johnson, *USGS Professional Paper 1180*, 1980.

Line Segment	100–500'	500–1000'	1000–1500'	1500–2000'	2000–2500'
Date	*Mean elevation of line segment*				
1950 Sept. 14	6,460.1	6,523.3	6,564.8	—	—
1951	—	—	—	—	—
1952 Aug. 22	6,460.3	6,522.6	6,563.8	6,604.8	—
1953 Sept. 4	6,458.4	6,519.5	—	—	—
1954 Sept. 27	6,459.5	6,522.0	6,564.4	—	—
1955 Sept. 6	6,460.6	6,521.8	6,563.9	—	—
1956 Aug. 30	6,461.7	6,521.6	6,563.8	6,604.6	6,659.9
1957 Sept. 10	6,456.6	6,517.9	6,560.6	6,600.9	6,654.9
1958 Sept. 15	6,446.4	6,509.8	6,551.6	6,591.2	6,642.7
1959 Sept. 12	6,449.9	6,513.6	6,555.7	6,597.1	6,649.5
1960 Sept. 6	6,449.9	6,514.1	6,555.5	6,594.7	6,646.7
1961	—	—	—	—	—
1962 Sept. 2	6,442.8	6,506.4	6,547.4	6,596.9	6,640.7
1963 Sept. 12	6,437.5	6,499.2	6,541.8	6,582.4	6,634.4
1964 Sept. 15	6,436.5	6,499.1	6,541.1	6,581.4	—
1965 Sept. 6	6,436.4	6,499.6	6,540.9	6,580.2	6,635.0
1966 Aug. 17	6,437.6	6,501.5	6,543.1	6,582.3	6,637.7
1967	—	—	—	—	—
1968 Aug. 28	6,432.1	6,498.8	6,539.5	6,578.8	6,633.9
1969 Aug. 27	6,428.2	6,493.3	6,534.2	6,574.2	6,628.3

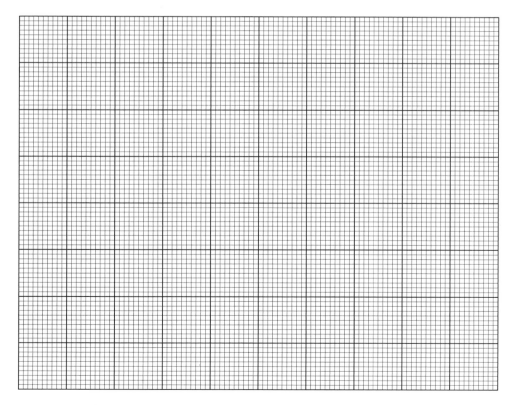

FIGURE 10.15
Graph paper for Problem 8b.

CHAPTER 11

Sea Coasts

Materials Needed

- Pencil and eraser
- Ruler
- Stereoscope (provided by instructor)

Introduction

The land along the coasts is subject to relentless pounding by waves. These waves move a great deal of sand and other sediment that can maintain beaches, make them disappear, move them inland, or push them across the mouths of bays and harbors. Where waves hit coastal cliffs, they focus their energy on the irregularities that jut into the sea. These promontories gradually wear away and collapse into the sea, and the sediment they produce is moved along to help fill in nearby bays and estuaries. Since over 75 percent of the U.S. population lives near the seacoasts, it is worth understanding how waves, longshore currents, and changes in sea level affect the coasts. In this lab, you will learn to recognize common coastal features on maps and in air photos. You will learn to use them to interpret the processes that have affected, are affecting, and will affect a given coast.

Waves, generated by wind blowing across the water, play the greatest role in shaping the coast. Figure 11.1 illustrates a wave and shows how its passage causes movement of water. Note that while the crest of a wave travels across the water, the water itself simply moves in a circular path. Water motion decreases to zero at the **wave base,** a depth approximately equal to half the wavelength.

As waves approach the shore, they begin to interact with the bottom, their shapes change, and the pattern of water movement changes, as shown in Figure 11.2. When the water depth becomes less than wave base, waves begin to erode and transport fine-grained bottom sediment. In the shallow **surf zone,** the wave shape breaks down, and water motion becomes turbulent; it is here that most work is accomplished. Sand and pebbles are thrown into suspension by the turbulent surf and are in near-constant motion.

Waves nearly always approach a coast with their crests at an acute angle to the shoreline. As the waves begin to interact with the bottom, they slow down and gradually bend, or **refract,** and become more parallel to the coast, as shown in Figure 11.3. In spite of refraction, however, the waves still hit the shore at an angle. Although most of the wave energy is directed at the beach, some is parallel to the beach. This produces a **longshore current,** which flows along the coast in the direction in which it is being pushed by the waves (Fig. 11.3). Sediment in the surf zone is carried by the current in a process known as **longshore drift.**

Wind energy is thus transferred through wave action to the surf zone, where erosion, transportation, and deposition combine to modify the coastline. We look first at common landform features that result from coastal erosion, then turn to those features formed by deposition.

Erosion

All coasts undergo erosion, but the intensity differs among coasts. The evidence for active erosion is best seen on rocky coasts. Figure 11.4 shows waves being refracted by the shallows around a **headland,** which is a rocky point of land between bays. Notice how the waves bend to focus their destructive energy directly on the headland. The concentration of erosive activity just above and below sea level creates steep **wave-cut cliffs.** These are progressively undercut and collapse into the water. In time, this process creates a gently sloping erosional surface just below sea level called a **wave-cut platform.** More resistant portions of rock temporarily left standing offshore on the wave-cut platform are **sea stacks.**

Deposition

While headlands erode because wave energy focuses on them, deposition takes place in the bays because wave energy dissipates there. Deposition results in the formation of a **beach,** commonly made of sand, pebbles, or cobbles eroded from the headlands, as well as material carried to the coast by rivers. In time, the combination of erosion and deposition reduces the irregularities of the coastline.

Longshore drift aids in straightening out coastlines. When a longshore current

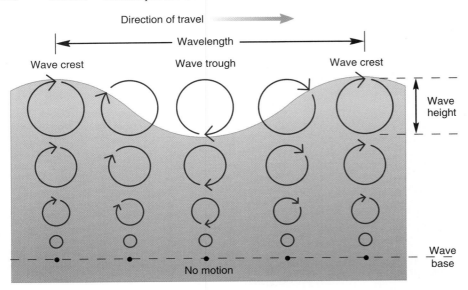

FIGURE 11.1

Cross section of wave, showing how water motion decreases to zero at wave base (a depth approximately equal to half the wavelength).

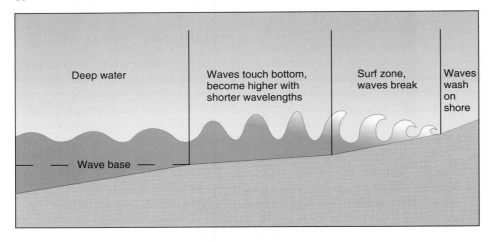

FIGURE 11.2

When waves enter water shallower than the wave base, they change form until, in the surf zone, they break, and water motion becomes turbulent.

passes a bend in the coast and enters deeper water, it slows, and some of its load is deposited. A common feature formed by this process is a **spit,** an emergent ridge of sand growing in the direction of longshore drift, as shown on Figure 11.5. A spit extending completely across the mouth of a bay is a **baymouth bar. Tombolos** are sand ridges connecting islands to the shore (Fig. 11.5). They develop because the island blocks and refracts the waves, interrupting the longshore current. The refracted waves push the sand behind the island where it is deposited.

Rivers provide most of the sediment for beaches and longshore drift. If longshore currents are strong, most of the sedi-

ment at the mouth of a river is swept away. If longshore currents are weak, or if the sediment load of a river is excessive, the sediment is deposited at the mouth of the river as a **delta** (Fig. 11.5).

Low-lying coasts, such as the southeastern U.S. coast from New York to Mexico, may have long, narrow, sandbar islands paralleling the coast. These **barrier islands** (Fig. 11.6) are 2 to 5 km wide and 10 to 100 km long. They are separated from the mainland by a *lagoon* some 3 to 30 km wide. Barrier islands slowly migrate landward due to big storm waves washing sand from the ocean side to the lagoon side. This process creates **overwash fans,** lobes of

sand that breach the vegetated dunes and extend toward the lagoon. Gaps between islands, known as **tidal inlets,** develop strong tidal currents as the tide rises and falls. Sediment transported by these currents is deposited as **tidal deltas** both landward and seaward of the inlets. Longshore currents often push spits into tidal inlets. These spits generally curve toward the lagoon due to interactions between waves, longshore currents, and tidal currents. As the spit closes off one side of the inlet, tidal currents erode the other side to keep the inlet open. This causes tidal inlets to migrate along barrier islands in the direction of longshore drift. Both wave action and inlet migration can dramatically change a barrier island within even a lifetime. Despite this, several major cities (including Atlantic City, New Jersey; Miami Beach, Florida; and Galveston, Texas) and hordes of resort developments are located on barrier islands.

Human Interaction

People living along coasts, and especially on barrier islands, commonly do not understand the power and complexity of the natural processes of coastal erosion and deposition. If their part of the beach is eroding, they may try to protect their property from encroaching sea waves by erecting a concrete **seawall.** Unfortunately, reflection of wave energy off this seawall further enhances erosion of the beach, and a narrower beach allows waves to more forcefully pound the seawall. Eventually the seawall fails and a larger, uglier, more expensive seawall must be built.

Another approach is to build **groins,** which are low walls built perpendicular to the beach to interrupt the longshore current (Fig. 11.7). Although sand is deposited on the upstream side of the groins, it is eroded away on the downstream side. Thus, some property owners have a wider beach, others have a narrower beach, and everyone gets to look at ugly groins cutting up the beach. Hard structures should in general be avoided for beach protection because they have unintended negative consequences, are expensive to maintain, and are politically impossible to remove once in place.

Another popular beach-saving strategy is to dredge sand from a tidal inlet or offshore area and, at a cost of tens of millions

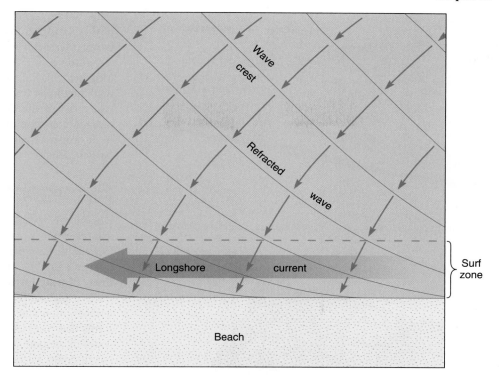

FIGURE 11.3

Map view showing refraction (bending) of waves as they approach shore. Solid lines represent wave crests, and arrows indicate wave movement. As the water is pushed against the shore, a longshore current develops in the surf zone. The longshore current carries sediment with it in the surf zone, a process known as longshore drift.

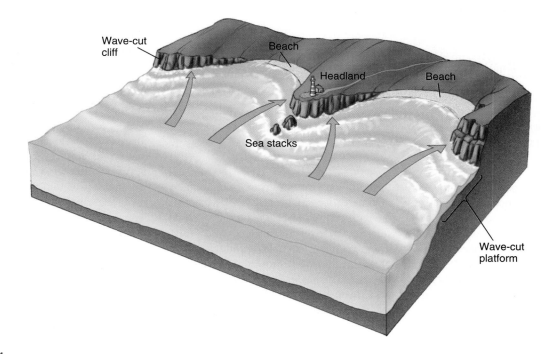

FIGURE 11.4

Waves approaching an irregular coastline are refracted so that energy is concentrated at headlands. Erosion results in the formation of wave-cut cliffs, sea stacks, and a wave-cut platform.

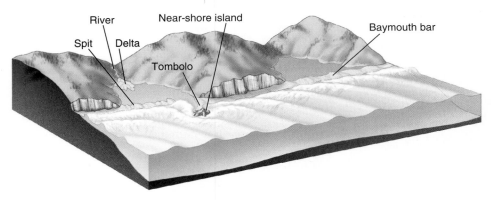

FIGURE 11.5
Longshore drift is from left to right, as indicated by the spit. Wave refraction around the near-shore island causes the tombolo to form.

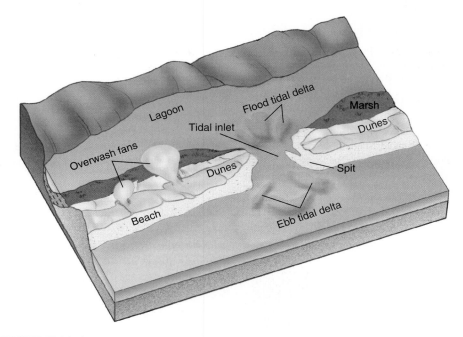

FIGURE 11.6
Features of barrier islands include beach, dunes, overwash fans, marsh, lagoon, tidal inlet, and tidal deltas (usually submerged).

of taxpayer dollars, pump it onto the beach. Of course, the processes that cause the beach to naturally move have not changed, and the precious sand from such "renourishment" projects frequently disappears within a few short years. It is difficult to reconcile the natural mobility of the beach with immobile buildings and fixed property lines.

Submergent Coasts

The position of shorelines has fluctuated throughout geologic time. Over the past three million years, dramatic changes in the volume of ice stored in the polar ice caps have caused sea level around the world to fluctuate by over 100 meters (>325 feet). There are many examples around the world of coasts that were drowned (submerged) by the most recent sea level rise, which started some 20,000 years ago as the great ice sheets began melting. Shorelines can also move in response to tectonic and other forces that cause the edges of continents to rise and fall relative to sea level.

The characteristics of **submergent coasts** depend on the nature of the land-

scape before submergence. If the topography was rugged, the submerged coastline is irregular, with such features as **estuaries** (drowned river valleys) or **fiords** (drowned glacial valleys) and islands. If the topography had low relief, such as the east coast of the United States, the submerged coastline may have large estuaries, extensive tidal flats, salt marshes, and barrier islands.

Emergent Coasts

Despite the recent sea level rise, tectonic and other forces have uplifted many coasts, especially along the tectonically active Pacific margins, to create **emergent coastlines. Marine terraces** are their most distinctive feature (Fig. 11.8). Many marine terraces represent normal wave-cut platforms that were later uplifted by brief episodes of tectonic activity. Some were also cut 125,000 years ago when global sea level was about 2 meters higher than today or at other times over the past few hundred thousand years when sea level was similarly high.

Recent and Future Sea Level Rise

The Earth's climate has warmed by about 0.6°C (1°F) over the past 100 years. This warming and its likely continuation will cause global sea level to rise. A rise of 20 cm over the past 100 years has already resulted from the accelerated melting of mountain glaciers and continental ice sheets and the warming and consequent thermal expansion of the oceans. Unfortunately, even such small increases in sea level are likely to significantly increase coastal erosion and flooding. Along coasts with sea cliffs, a higher sea level allows waves to carry more of their energy over the beach and directly into the sea cliffs. With barrier island beaches, a raised sea level allows more storm waves to breach the vegetated sand dunes and wash sand into the lagoons (Fig. 11.6). The net result is accelerated erosion of both coastal sea cliffs and beaches. In addition, storm surges during hurricanes and other major storms are more likely to flood low-lying areas and destroy property. Global warming affects much more than just the weather!

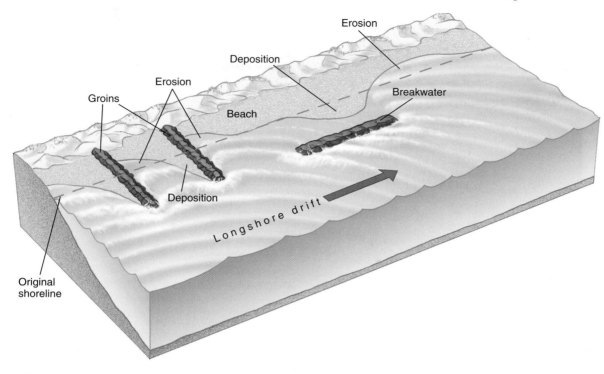

FIGURE 11.7
Groins and breakwaters interrupt longshore drift, causing deposition *and* erosion.

FIGURE 11.8
The gently sloping surface near Pacific Valley, California, was a submerged wave-cut platform that has been uplifted to form a marine terrace. Old sea stacks are visible on the terrace. Although now vegetated, such terraces often preserve well-rounded beach gravel and marine shells (clams, snails, etc.) as evidence of their former submergence.

Hands-On Applications

Coastlines are among the most dynamic environments in which we live, so it is important that we thoroughly understand coastal processes. Geologists learn to use the past and the present to predict the future. This lab demonstrates how careful observation of past results and present processes, together with the application of general principles, leads to an understanding of the coastal environment.

Objectives

If you are assigned all the problems, you should be able to:

1. Illustrate the refraction of waves as they strike a smooth or irregular coastline.

2. Explain the origin of longshore currents and longshore drift.

3. Explain how wave refraction may determine where erosion and deposition take place along a coast.

4. Identify the following erosional features on a map or in a photograph: headland, wave-cut cliff, wave-cut platform, sea stack.

5. Identify the following depositional features on a map or in a photograph: beach, spit, baymouth bar, tombolo, delta, barrier island, and associated features such as lagoon, tidal inlet, tidal delta, tidal flat.

6. Determine the dominant direction of the longshore current from a map or photograph showing features such as spits or groins.

7. Based on a topographic map, identify a coast as rocky or sandy, and use evidence on the map to determine whether erosion or deposition is the dominant process.

8. Recognize submergent and emergent coasts and their distinctive features, such as estuaries, fiords, and marine terraces.

Problems

1. Figure 11.9 illustrates the fragility of barrier islands with a sequence of aerial photographs. Shown in the photographs is part of Matagorda Island, a barrier island off the Texas coast in the Gulf of Mexico. Hint: Figures 11.5 and 11.6 may help you answer the following questions.

    a. Photo A, from 1943, shows a tidal inlet called Greens Bayou. Note the road to the left (southwest) of Greens Bayou. What is the minimum width of Greens Bayou? (The fractional scale is 1:10,200.)

    Based on the orientation of the waves in Photo A, determine the direction of the longshore current (right to left or left to right). What, if any, evidence is there to indicate whether this is the usual direction of longshore current?

A. 1943, scale 1:10,200

B. 1957, scale 1:25,400

C. 1961, scale 1:18,500

FIGURE 11.9
Sequence of three sets of stereophotographs of Greens Bayou, Matagorda Island, Texas, taken in: A. 1943 (scale 1:10,200); B. 1957 (scale 1:25,400); and C. 1961, after Hurricane Carla (scale 1:18,500). The white arrows mark an identical feature on A and B.

b. Photo B shows the same area in 1957. The coastal road and white arrow should help you to orient yourself with respect to Photo A.

What is the minimum width of Greens Bayou now? Measure the width just to the upper right of the end of the road where the two segments of the island are closest together. The fractional scale is 1:25,400.

Where have erosion and deposition occurred?

You have seen how the width of a tidal inlet can change, which is of some concern to boaters. How about the position of the tidal inlet? If you owned property on a barrier island, you would be more concerned about a moving inlet destroying your yard and home. The white arrows point at the same spot (probably a clump of vegetation) in photos A and B. Measure the distance in a line parallel to the coast between this clump of vegetation and the far side of the tidal inlet for each photo. Remember, the scales are different in A and B.

Distance in 1943:

Distance in 1957:

How far has the left-hand side of the inlet moved over these 14 years?

What is the origin of the set of curved features located to the right of the inlet in 1957?

As the tidal inlet has narrowed, the current velocities resulting from the tides moving in and out through the inlet have dropped. How might this be related to the large shallow-water sand deposit behind the inlet? What is this sand deposit called?

c.   The Texas coast is frequently hit by hurricanes; 46 struck during the twentieth century. On September 14, 1961, one of the largest hurricanes in recorded history, Hurricane Carla, hit Matagorda Island. The hurricane virtually destroyed the Matagorda Island Air Force Base and considerable erosion occurred along the island. Photo C shows Greens Bayou in 1961 after Carla struck. The coastal road is barely visible just behind the beach.

What is the minimum width of Greens Bayou in Photo C? The fractional scale is 1:18,500.

Comparing the width of the tidal inlet and the position of the coastal road with the earlier photos, describe the major changes that have taken place as a result of Carla.

Is a barrier island a good place to use your retirement savings to build a condo?

2.   Figure 11.10 is a portion of the Cayucos, California, 1:62,500 quadrangle, which is about 15 miles northwest of San Luis Obispo.

a.   Where would you go to see a rocky beach? A sandy beach? (Hint: See the table of topographic map symbols in Chapter 6.)

What coastal landform is represented by the small circular land areas just offshore 2 to 6 km northwest of Morro Beach (for example, Whale Rock)?

FIGURE 11.10

Portion of the Cayucos, California, 15-minute quadrangle, for Problem 2.

Scale 1:62,500
C.I. 50 feet

b.   What is the landform on the west side of Morro Bay, labeled Morro Bay State Park?

Morro Rock is shown as an island on older maps. What landform now connects Morro Rock to the mainland?

What happens to Morro Creek as it approaches the coast? Explain. How might this situation change during a period of abundant rainfall?

c.   What does the pattern of dots in Morro Bay indicate? (Hint: See the table of topographic map symbols in Chapter 6.)

What depositional feature is the swampy area on the east side of Morro Bay?

What does the future hold for Morro Bay?

d.   In what direction is the longshore drift in the area between the town of Morro Beach and the bottom of the map, as indicated by (1) the breakwater and (2) two different natural landforms (specify each)? What is the evidence?

3.   Figure 11.11 is a portion of the Boothbay, Maine, 1:62,500 quadrangle. This area, about 40 miles northeast of Portland, is made of metamorphic rocks whose resistant layers intersect the surface in generally north-south bands. The striking topography and extremely irregular coastline result from differential erosion of these rocks. Several of the long bays are named after the river that flows into them. Blue contours show water depth in feet.

a.   Is this a rocky coast or one dominated by sandy beaches? Cite evidence for your answer.

b.   Is erosion or deposition the dominant process along most of the coast? Cite evidence for your answer in terms of landforms and map symbols. Consider both on-land and offshore topography.

c.   Is this coast emergent or submergent? Cite evidence.

d.   Use geographic names to cite examples of the following features: wave-cut platform (identified by the symbol for "rock or coral reef"—the symbol given in Fig. 6.10 is very similar to the symbol used on the map, though it is not identical), sea stack (identified by the symbol for "rock, bare or awash"—see Fig. 6.10), headland, wave-cut cliff, estuary. Consider subaerial and submarine topography when searching for examples.

FIGURE 11.11
Portion of Boothbay, Maine, 15-minute quadrangle, for Problem 3.

e. If you have completed the chapter on glaciers, you may recognize landforms suggesting glacial erosion. In fact, continental ice sheets covered this area from roughly 25,000 to 11,000 years ago. During this time so much water was locked away in continental ice sheets that sea level fell approximately 120 meters and exposed the entire area seen in Figure 11.11. Identify one aspect of the topography suggesting that the area was once glaciated. What about the topography indicates the ice flow direction?

f. If climatic and sea-level conditions remain similar to those of the present, predict how this coastline might look in 10,000 or 20,000 years. Consider the effects of erosion and deposition by both marine and terrestrial processes. Sketch a possible new shoreline on the map, and clearly label those areas removed by erosion or built by deposition.

4. Go to *http://ceres.ca.gov/ceres/calweb/coastal/geography.html* (or link to it through *www.mhhe.com/jones6e*—see Preface), and answer the following questions about marine terraces and beaches along the California coast.

a. Marine terraces

What are marine terraces and where along the California coast are they seen?

How are marine terraces formed?

In some places, up to twenty marine terraces are present. What combination of events occurred that allowed so many wave-cut platforms to form, then rise above sea level to form terraces?

b. Beaches

From where does the sand in California beaches come?

What is *littoral drift?*

How is sand permanently lost from the beach system?

How have humans altered the natural beach processes and what effect might this have on the system?

In Greater Depth

5. Figure 11.12 is a portion of the Redondo Beach, California, quadrangle from 1951. Figure 11.13 is a 1990 stereopair of the same region. The area is about 25 miles south of Los Angeles proper.

    a. Is the coast rocky or sandy in this area? Cite your evidence.

    b. Draw a topographic profile between A and A' using the graph paper of Figure 11.14. Note that the contour intervals on land and offshore are different, and that on land, every *fourth* line is an index contour. Use every contour when preparing your profile or you will not see that the profile is steplike, with areas of gentler slope separated by areas with steeper slopes.

    c. Refer to your topographic profile. How did the gently sloped area immediately seaward of the coastal cliff form, and what landform name is given to such a feature?

    d. How do you think the other gently sloped areas on your topographic profile formed?

        How many of these areas are there, including the one just off the coast? Number them for reference purposes; assign "1" to the one just off the coast, "2" to the first one inland, and so forth.

        What is the approximate change in elevation between each of the flat areas? For example, between 1 and 2, 2 and 3, and so forth. Measure elevation differences between the top and bottom of each area with steep slope.

    e. Look at the stereopair in Figure 11.13. These photos were taken in 1990. Note the amount of development since 1951, and where it has occurred.

        Is erosion or deposition likely to be the dominant process along the coast? Circle those areas that ought to pay the highest rates for natural hazards insurance, and explain your answer.

    f. What does the topographic profile along A-A' suggest about the manner in which emergence took place? That is, was it gradual, episodic, or can't you tell? Explain. (Be careful; this is tricky, because two variables must be considered.)

    g. Assign relative ages to the flat areas on the profile. That is, which appears to be the youngest, which the oldest, and where in the sequence do the others fit? (Hint: Which have been subjected to the least on-land erosion, which the most, and which are intermediate?)

Scale 1:24,000
C.I. 25 feet

0   1/2   1 Mile

0   .25   .5   1 Kilometer

FIGURE 11.12
Portion of the 1951 Redondo Beach, California, 7½-minute quadrangle, for Problem 5.

FIGURE 11.13
Infrared stereopair of Palo Verdes Point area, California, 1990. Scale is 1:40,000. North is to the top. Compare with Figure 11.12 to see population changes from 1951 to 1990. Large red areas are golf courses.

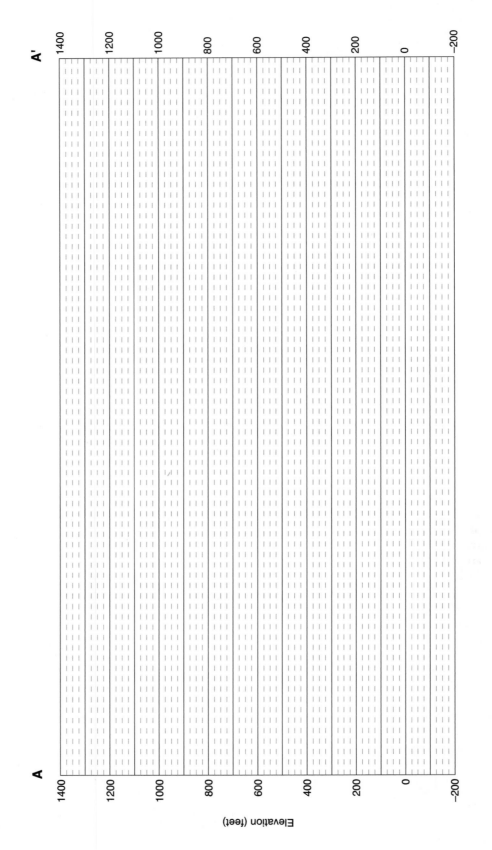

FIGURE 11.14
Graph paper for a topographic profile, Problem 5.

CHAPTER 12

Arid-Climate Landscapes

Materials Needed

- Pencil and eraser
- Colored pencils
- Metric ruler
- Calculator
- Stereoscope (provided by instructor)

Introduction

The bold, stark landscapes of arid and semiarid lands contrast with the rounded, more subtle forms of humid regions. Although running water is the principal agent of erosion in warm arid regions, it commonly is present only after infrequent cloudbursts. Flash floods remove the products of mechanical weathering and mass wasting and leave behind steep banks and cliffs. Plateaus, mesas, and buttes develop from deep erosion of horizontal layers of resistant bedrock. As erosion wears down mountains, alluvial fans form at canyon mouths and, with time, gently sloping erosional and depositional surfaces extend outward from the mountain front. Where strong winds blow and sand is available, dunes of several types may form. In this lab, you will learn to recognize and analyze common landforms that develop in warm arid climates, such as those found in the U.S. southwest. Arid landscapes also form in cold regions (e.g., the north slopes of Alaska and the dry valleys of Antarctica), but we do not consider them here.

Warm arid and semiarid regions (precipitation, <50 cm/year; see Fig. 8.1) are characterized by sparse vegetation, little water, thin soils, frequent strong winds, and sharp, angular landforms. Mechanical weathering predominates, and sedimentary particles tend to be coarser than in humid regions. Slopes typically are steeper, a requirement for the removal of coarser particles. Even rock such as limestone, which is readily attacked by chemical weathering in humid climates, forms steep cliffs in dry climates.

In spite of the aridity, water is the principal agent of erosion in most deserts. Rain commonly comes as downpours that quickly fill streams and cause flash floods. These floods cause extensive erosion, and sediment is quickly transported down normally dry stream beds and out into valleys before it is deposited. When sediment-choked streams spill out into a valley, they spread out to form numerous interwoven shallow channels called **braided streams** (Fig. 12.1). These braided streams rapidly deposit their sediment to form deposits of sand and gravel. Individual channels have steep, easily eroded banks. Each new flood tends to rework the older sediment and may carve a new main channel among the many secondary ones.

Wind is also an important agent of erosion and transportation in deserts, but it does not move as much material as water does. In fact, large, wind-carved rock landforms are unusual. Sand dunes, however, are common, and demonstrate that wind is an active agent of sediment transport and deposition.

Erosional Landforms

While most landforms in arid and semiarid regions are shaped by running water, the streams in arid regions are different from those in humid climates. Most streams are **intermittent,** carrying water only part of the year. Many, whether intermittent or **perennial,** have braided channels (Fig. 12.1). Deep canyons in dry climates may have near-vertical walls, and box-like cross profiles.

Where rivers traverse and cut into horizontal layers of sedimentary or volcanic rocks, *cliff-and-bench* topography, like that in Figure 12.2, develops. The more resistant layers of sandstone, limestone, or volcanic rock form steep cliffs where erosion has cut through them; flat benches develop on their upper surfaces. Less resistant layers of shale or soft volcanic rock form gentler slopes between the benches. The cliff-and-bench topography is developed in a spectacular way on the *Colorado Plateau* of the western United States. This region, in eastern Utah, western Colorado, northwestern New Mexico, and northern Arizona, has sixteen national parks or monuments featuring this scenery, including Grand Canyon and Canyonlands.

Plateaus are comparatively flat upland areas that form in regions with nearly horizontal rock layers. Figure 12.3 illustrates several common features developed by the

FIGURE 12.1
Upstream view of a dry stream bed in Death Valley National Park, California, with braided channels and easily eroded banks.

FIGURE 12.2
Cliff-and-bench topography in Canyonlands National Park, Utah, along the Green River.

dissection of a plateau. As stream valleys enlarge by lateral and headward erosion, parts of the plateau may be cut off to form mesas. A **mesa** (table, in Spanish) is a generally flat-topped area, bounded by cliffs, that is wider (commonly by quite a bit) than it is high. **Buttes** are smaller, flat-topped landforms isolated by erosion. They are more or less as wide as they are high. **Monuments** (or spires) are slender features that are much higher than they are wide.

Where the bedrock does not consist of horizontal layers of unequal resistance, different erosional landforms develop. In the *Basin and Range* area of the southwestern United States, rugged mountain ranges are separated by wide valleys with flat floors. The mountains were uplifted along steep faults. Although some faults are still active and are readily discernible, others have been obscured by erosion and deposition. Figure 12.4 shows a profile from a mountain into the adjoining valley. A sharp break in slope at the edge of the mountain marks the *mountain front*. The gentle valleyward slope from the mountain front is the **piedmont**. It consists of two parts, the *bajada* and the *pediment*. The bajada is formed by deposition, as discussed in the next section. A **pediment** is a gently sloping, erosional surface. It cuts across bedrock, and typically has a thin veneer of sediment on its surface. It is generally thought that a pediment develops as the steep slope at the mountain front retreats under the attack of erosion. The erosional debris is transported

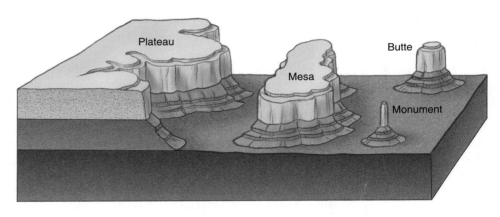

FIGURE 12.3
Characteristic erosional landforms in arid or semiarid areas with horizontal rock layers. Retreat of a plateau leaves mesas, buttes, and monuments as remnants.

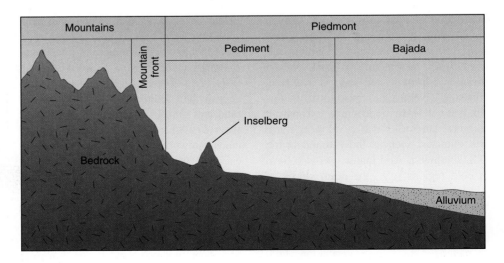

FIGURE 12.4
Profile from mountain into valley, showing the mountain front and the piedmont, with pediment, inselberg, and bajada.

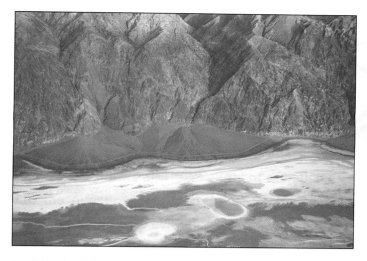

FIGURE 12.5

Two alluvial fans occur along the base of the Black Mountains in Death Valley National Park, California. There are two wet playa lakes in the foreground. The white areas are evaporite minerals that were deposited as more extensive playas dried up.

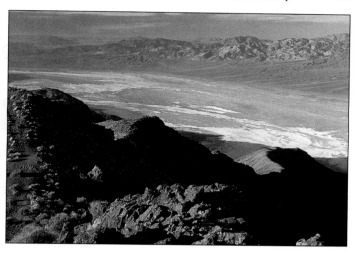

FIGURE 12.6

The dark area in front of the background mountains is a piedmont. The upper part of the dark area is the pediment and the lower part is the bajada, but you would need to be on the ground to map the boundary between them. White areas in the middle of the valley are evaporite deposits. Panamint Range, Death Valley National Park, California.

across the pediment to the bajada. Erosional remnants, or **inselbergs,** form isolated hills on the pediment.

Depositional Landforms

Water Deposits

Most sediment in arid lands is transported by intermittent streams. In mountainous areas, these streams cut and occupy rock-walled channels, and sediment transport is confined within the rock walls. As the canyons empty into a valley, the stream is no longer confined; it spreads out, losing its capacity to carry sediment. The resulting series of braided stream deposits at the canyon mouth eventually builds up a large **alluvial fan,** like those shown in Figure 12.5. With time, alluvial fans from adjacent canyons coalesce to form a **bajada,** the gently sloping apron of sediment along the mountain front in Figure 12.6. If a pediment is present, the bajada lies basin-ward of the pediment, and together, they form the piedmont at the base of the mountain (see Fig. 12.4).

When an intense rainstorm occurs in arid areas, water runs out of the highlands, down canyons, and across the piedmont. If a pediment exists, much of the water washes across it to the bajada. As it crosses

the bajada, a substantial amount of water may soak into the alluvium and become part of the groundwater. But, if it is a real downpour, much water flows out into the valley. If there is no through-going drainage, water may pond to form temporary lakes called **playas** (Fig. 12.5). Evaporation of water from playas results in the deposition of evaporite minerals, such as gypsum or halite.

Wind Deposits

The most obvious wind deposits are sand dunes. They develop from the accumulation of sand-size grains that have been transported by the wind. Grains move by rolling and hopping, and rarely get more than a meter above the surface; many are set in motion by impact from other grains.

Initial deposition generally is caused by an obstacle or irregularity on the surface. Wind velocity decreases after it blows over an obstacle, and some of the sand is deposited on the downwind side; continued deposition produces a dune. Figure 12.7 is a cross section of a dune, showing how a steep, downwind face develops as sand rolls and hops up the windward side and tumbles and slips down the leeward side, forming cross-strata (see Fig. 4.5). As this process continues, the dune migrates downwind.

Differently shaped dunes develop under different conditions. The principal con-

trols are wind velocity, vegetation, and sand supply.

Transverse dunes (Fig. 12.8A) are wave-like ridges of sand perpendicular to the predominant wind direction. They form in areas of abundant sand supply, scarce vegetation, and moderate, unidirectional winds. They are found not only in deserts, but also commonly behind beaches and on barrier islands.

Longitudinal dunes (Fig. 12.8B) are long, narrow ridges of sand parallel to the prevailing wind direction. They form where sand supply and vegetation are meager and winds are strong. Although they parallel the general wind direction, the slip face varies from one side to the other along the axis of the dune, reflecting variations in wind direction.

Parabolic dunes (Fig. 12.8C) are crescent shaped, with the steep slip face on the convex side so that the horns point upwind. They typically develop where vegetation is available to anchor the horns. In many, the area between the horns is a **blowout,** a small depression excavated by the wind.

Barchan dunes (Fig. 12.8D) are crescent shaped, like parabolic dunes, but the steep slip face is on the concave side so that the horns of the crescent taper and point downwind. Barchans develop best on barren desert floors where prevailing wind direction is constant, vegetation is scarce,

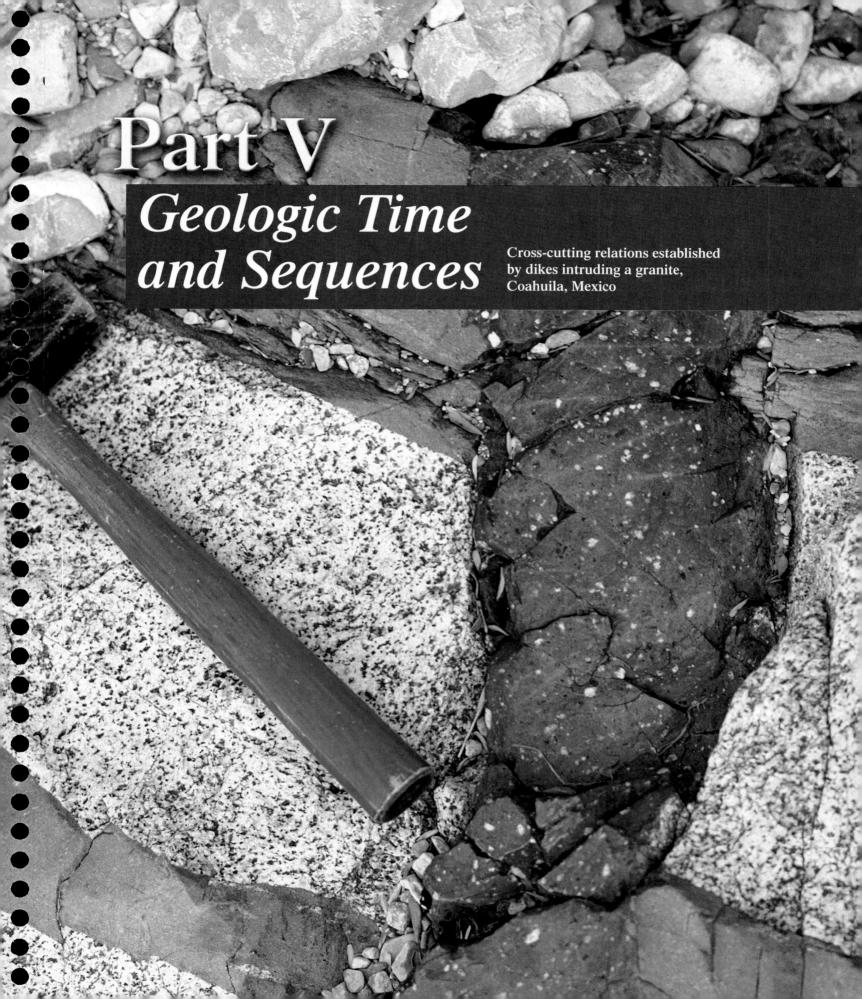

Part V
Geologic Time and Sequences

Cross-cutting relations established by dikes intruding a granite, Coahuila, Mexico

CHAPTER 13

Geologic Age

Materials Needed

- Pencil and eraser
- Metric ruler
- Calculator

Introduction

Rocks exposed at the Earth's surface often have long and complex histories. To understand the formation of these rocks, and thus to unravel the history of the Earth, it is necessary for geologists to use rigorous techniques to figure out not only what but also when things happened. For centuries, geologists have determined the *relative* ages of different events using simple observational techniques: this sandstone was deposited first, then came a limestone, and finally a granite intruded both. Only since World War II have geologists developed the sophisticated methods required to determine numerical ages. Thus, a sandstone was deposited 250 million years ago, a limestone 245 million years ago, and a granite intruded both 40 million years ago. The problems in this chapter illustrate the techniques used to determine both relative and numerical ages.

Geologists working in mountain ranges are regularly confronted with the complexity of the Earth's past. Instead of seeing merely horizontal layers of sedimentary rock, we often see sedimentary layers that are folded or steeply tilted (Fig. 13.1A, B). Other layers may be abruptly offset by fractures called faults (Fig. 13.1B). And sometimes igneous rocks have clearly intruded sedimentary rocks, or sedimentary rocks were deposited on top of older, cooled igneous rocks (Fig. 13.1C, D). These geologic relationships record the forces and events that help shape the Earth.

It was in the late 1700s that James Hutton, the father of modern geology, realized that the events recorded in the rock record must have taken a very long time to unfold. He and his contemporaries used careful field observations and scientific principles to recognize many different types of events, including deposition and burial of sediment, igneous activity, rock deformation, and uplift and erosion of preexisting rocks. They also were able to arrange these events in a *relative* sequence from oldest to youngest. However, they were unable to assign exact dates to anything that occurred before the beginning of recorded history. It was like knowing only that you are younger than your mother, and that she is younger than your grandfather, but not knowing anyone's actual age.

Geologists now are able to determine quite precisely the dates, or **numerical geologic ages** (also known as *absolute geologic ages*), of many types of geologic events. Determining a numerical geologic age is complex and expensive, so numerical ages are not always readily available. However, many thousands of numerical ages have been determined so that if the *relative* geologic age is known, an estimate of the numerical age can be made. Both types of ages, relative and numerical, are important pieces of information for interpreting and understanding the geology of an area, unraveling the complexities of past tectonic-plate movements and interactions, and reconstructing the history of climate change and the long pageant of prehistoric life.

Relative Geologic Time

General Concept

Although it is important to know the numerical age of a particular geologic event, it is equally important to be able to indicate

A. Folded rocks

B. Tilted and faulted rocks

C. Igneous dike that intruded older sediments

D. Igneous rock upon which sediments were deposited

FIGURE 13.1

Perspective sketches of some geologic structures that might be observed in roadcuts or cliff faces. A. Folded rocks. B. Tilted rocks with a fault offsetting the sedimentary layers. C. Intrusive igneous rock, in this case a dike (a tabular body filling an opened fracture). The crosshatches along the igneous/sedimentary rock contact indicate a contact metamorphic rock called hornfels. D. Intrusive igneous rock, exposed to erosion and later buried by sediment. No contact metamorphism is apparent.

whether that event occurred before or after another event. This is the basis for a **relative geologic age** the age of one event relative to another.

A description of the geology of an area includes a list of the geologic events that took place, in the sequence in which they occurred, from oldest to youngest. Geologic events that are commonly described include deposition of sedimentary units, extrusion or intrusion of igneous rocks, metamorphism, folding, faulting, uplift, and erosion.

To understand how a sequence of events is determined, consider Figure 13.2, a sketch of a roadcut in a mountainous area. The sketch shows three inclined layers of sedimentary rock sandstone, shale, and limestone intruded by a granite dike. What geologic events must have occurred

to produce what is seen in the roadcut, and in what order did they occur?

Start with the simplest, the dike. Because it cuts (intrudes) all three sedimentary layers, it must be younger than all of them.

Next, what are the relative ages of the sedimentary layers? The one on the bottom must have been deposited first (assuming the layers are not upside down, a possibility that is considered later), and the one on top last.

So far we have the following sequence:

intrusion of granite dike **(youngest)**

deposition of limestone

deposition of shale

deposition of sandstone **(oldest)**

But there is more. Was the original sediment of the sedimentary rocks deposited as inclined layers? Not likely. The layers must have been tilted or folded after deposition.

And the dike was it intruded before or after the beds were tilted? Can you tell? Not without more information, of a kind that is too detailed to discuss here. So we must be satisfied with two possibilities for now; the dike could be older or younger than the tilting event.

What about the present-day surface, at the top of the roadcut? It is still forming, by erosion. Because the original sediments must have accumulated in a low place, and because they must have been buried under younger sediment in order to become lithified, the erosion must have been accompanied by uplift of the whole area.

FIGURE 13.2

View of rocks exposed in a vertical roadcut in a mountainous area. Sedimentary rocks are limestone (brick pattern), shale (parallel dashes), and sandstone (dots); granite dike shown with random dashes. What sequence of events led to what we see?

Now we can list the entire sequence in order of relative age:

uplift and erosion **(youngest)**

intrusion of granite dike (or tilting)

tilting of beds (or granite dike)

deposition of limestone

deposition of shale

deposition of sandstone **(oldest)**

As you can see, there is nothing magical or mystical about the way in which this sequence was determined. Logic and a few basic principles are all that's needed.

Basic Principles

From the preceding example, you can see that several fundamental principles exist to help interpret the relative time relations among rocks. Among the more useful are the following, the first three of which were formulated by Nicholas Steno in the late 17th century.

Steno's Principle of Original Horizontality (Fig. 13.3): Sediments are deposited in horizontal or near-horizontal layers. Therefore, non-horizontal layers have generally been folded or tilted from their original positions.

Steno's Principle of Superposition (Fig. 13.3): In any succession of sedimentary rock layers lying in their original horizontal position, the rocks at the bottom of the sequence are older than those lying above.

Steno's Principle of Original Lateral Continuity (Fig. 13.3): Sediments are deposited in layers that continue laterally in all directions until they thin out as a result of nondeposition, or until they reach the edge of the basin in which they are deposited. This means that if you find a layer that abruptly ends, something has cut this layer after it was deposited. Faults, dikes, and erosion can all truncate otherwise laterally continuous layers.

Principle of Cross-Cutting Relations (Fig. 13.4): Any geologic feature (intrusive igneous rock, fault, fracture, erosion surface, rock layer) is younger than any feature that it cuts.

Principle of Inclusions (Fig. 13.5): An inclusion in a rock is older than the rock containing it. Examples of inclusions are pebbles, cobbles, or boulders in a conglomerate, or xenoliths (pieces of other rocks) in igneous rocks.

Relative Ages Based on Fossils

These basic principles establish age relations among rocks that occur together in a local area. William Smith, working in the 1790s,

FIGURE 13.3

Sedimentary rocks viewed from Dead Horse Point State Park, Utah, illustrate the principles of Original Horizontality (sediment is deposited in horizontal layers), Superposition (younger beds are deposited on older beds), and Original Lateral Continuity (sedimentary layers do not abruptly end unless they hit the edge of a basin or are cut by younger features).

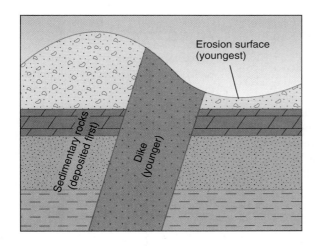

FIGURE 13.4

Principle of Cross-Cutting Relations. This cross section shows a dike cutting preexisting layers of sedimentary rock; the dike is younger than the rocks it cuts. The erosion surface cuts both the sedimentary rocks and the dike, so it is the youngest.

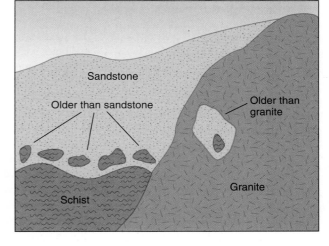

FIGURE 13.5

Principle of Inclusions. This cross section shows boulders of schist at the base of a sandstone; the boulders must be older than the sandstone containing them. Similarly, the xenolith of sandstone in the granite must be older than the granite.

examined the nicely layered sedimentary rocks and their fossils across England. (**Fossils** are the preserved remains, traces, or imprints of ancient plants and animals.) Employing the Principle of Superposition, he collected fossils from successive layers and discovered that he could use the fossils to determine the age-equivalence of widely separated sedimentary units, much as archaeologists recognize different historical periods in their excavations based on distinctive coins or pottery. Smith's work helped establish two important principles:

Principle of Fossil Succession: Fossil organisms succeed one another in time in a definite and recognizable order. Each distinct organism existed for a specific interval of time and not at older or younger times. The fossils in a sedimentary unit therefore can define a specific, unique interval of geological time.

Principle of Fossil Assemblages: Characteristic groups of fossil organisms also define unique geologic ages.

Fossils are an exceptionally useful means of determining relative time because they establish age relationships among widely separated rocks and because the sequence of fossil organisms is known over a very long interval of geologic time. The dinosaur *Tyrannosaurus rex,* for example, lived for a relatively short period of time. Any rocks found on any continent that contain its bones must date to this same short interval of time. Because the position of *T. rex* is known relative to the rest of the long fossil record, these rocks are securely located within all of geologic time.

Testing Hypotheses

In the example in which the sequence of events illustrated in Figure 13.2 was hypothesized, some assumptions were made that should not have been made without further observations.

The first assumption, a very reasonable one, was that the dike cut the layers of sedimentary rock. There is a *very* remote possibility that the dike was there first, standing as an inclined rock wall, while sediments were deposited on both sides of it. How would you tell? One way is to carefully examine the sedimentary rocks adjacent to the contact with the dike. For example, were they metamorphosed by the dike? If so, the dike must be younger. Do they contain any inclusions of the dike, such as pebbles, that might have been eroded from it? If so, the dike is older. Are there any small fingers of granite extending from the dike into the sedimentary rock that might have squeezed into weak places during intrusion of magma? If so, the dike is younger. Notice that the sedimentary layers cannot be projected straight across the dike but appear to be offset. Why? Because when the magma was intruded, it forced the rocks apart at 90° to the dike margins.

A second assumption was that the layers, though not horizontal, have not been completely overturned. If they have, then the limestone is the oldest, and the sandstone the youngest. How can you tell? One way is to look for sedimentary structures that allow you to tell top from bottom, up from down. Some useful sedimentary structures, shown in Chapter 4, are:

Cross-stratification (see Fig. 4.5)—cross beds are commonly cut off on the top of the bed and become parallel to adjacent layers on the bottom.

Oscillation ripple marks (see Fig. 4.4B)—symmetrical, wave-like features whose crests point to the top of the bed.

Graded beds (see Fig. 4.6)—grain size commonly becomes progressively finer upward.

Mud cracks (see Fig. 4.7)—in cross section, wider at the top than at the bottom.

Unconformities

The last event in the example shown in Figure 13.2 is erosion. What if sea level rose, or the land surface fell, and that erosion surface eventually was buried under younger sedimentary rock? The erosion surface would then be an **unconformity,** a surface that represents a substantial gap in the geologic record. It may be an ancient erosion surface, or it may be a surface on which neither erosion nor deposition occurred for a long period of time. If it is an erosion surface, it is recognizable because it cross-cuts older rocks, and its relative age can be determined by the Principle of Cross-Cutting Relations. If neither erosion nor deposition occurred, the unconformity may be difficult to recognize without studying fossils and applying the Principle of Fossil Succession. At any rate, there is no rock record of the time interval between the underlying rocks and those deposited on the unconformity.

Unconformities are of three principal types, each of which reflects distinct geologic events (Fig. 13.6). Figure 13.6A shows an **angular unconformity,** an erosion surface separating rocks whose layers are not parallel. Layers above and below meet at an angle. This implies that the underlying rocks were folded or tilted *and* eroded before the overlying layers were deposited. A **disconformity** is either an erosion surface or a surface of nondeposition separating rocks whose layers are parallel. An erosion surface is uneven and cuts layers of underlying rocks (Fig. 13.6B); a surface of non-deposition parallels underlying rock layers   (Fig. 13.6C). A **nonconformity** (Fig. 13.6D) is an erosion surface separating sedimentary rocks from older plutonic or massive metamorphic rocks (that is, crystalline rocks that are not layered).

Numerical Geologic Time

Numerical ages or dates can be determined in several ways. For example, you can tell how old a tree is by counting the number of growth rings. Or you can determine how long some glacial meltwater lakes existed by counting the number of varves—annually deposited sets of layers—present in the lake sediment. For obvious reasons, these and similar methods have

limited applicability. Radioactivity, however, provides a much more widely applicable method for determining numerical dates.

Radioactivity

Some isotopes of elements are **radioactive;** that is, their nuclei spontaneously break down or decay. In these reactions, a radioactive isotope, or **parent,** decays to form a different isotope, the **daughter.**

Isotopes are varieties of an element containing different numbers of neutrons in their nuclei. For example, the element uranium has two common radioactive isotopes, ^{235}U and ^{238}U. The superscripts 235 and 238 are the atomic masses or **mass numbers** (number of protons plus neutrons in the nucleus) of the isotopes. Each isotope has 92 protons in its nucleus—92 is the **atomic number** of uranium—but ^{235}U has 143 neutrons ($235 - 92 = 143$), whereas ^{238}U has 146 neutrons.

Ages from Isotopes

Laboratory measurements show that each radioactive isotope decays at a constant rate that is not affected by geologic age or surrounding temperature, pressure, or chemical conditions. The constant decay rate of each isotope enables us to calculate the numerical age of geologic samples. The decay rate is often described in terms of a **half-life,** which is the time required for half the atoms of any starting mass of a radioactive isotope to decay away. After one half-life, half of the parent atoms are gone, having produced an equal number of daughter atoms. After two half-lives, only half of one-half, or one-quarter, of the original parent remains; three-quarters of the parent has converted to daughter, so there are three times more atoms of daughter than parent.

The age of a mineral is determined from the relative amounts of parent and daughter isotopes it contains. The relative amounts of parent and daughter atoms are measured with an instrument called a mass spectrometer. If one knows the relative number of atoms of daughter and parent isotopes that are present, the ratio of these two combined with the *decay constant* can be used to determine the age. The **decay constant,** which is directly related to the half-life, describes the proportion of a starting mass of a given isotope that will decay

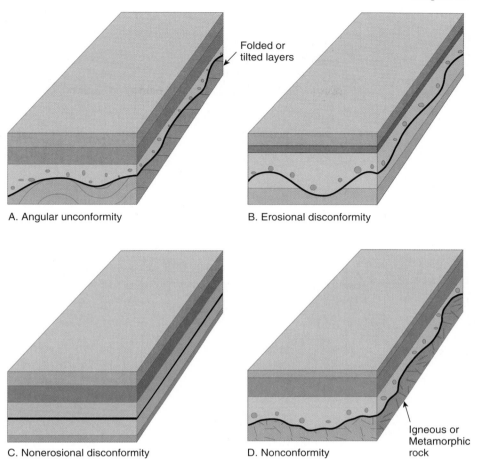

Folded or tilted layers

A. Angular unconformity

B. Erosional disconformity

C. Nonerosional disconformity

D. Nonconformity

Igneous or Metamorphic rock

FIGURE 13.6

Three-dimensional diagrams of: A. angular unconformity; B. erosional disconformity; C. nonerosional disconformity; D. nonconformity. The unconformity is an ancient erosion surface in A, B, and D, but in C it is a surface on which no erosion or deposition occurred for a long period of time.

away in a year. The age can be determined graphically or mathematically, as illustrated by the problems at the end of the chapter.

The age is calculated mathematically using the following equation:

$$t = ln(N_D/N_P + 1)/\lambda \quad \text{Eq. 13.1}$$

where t is time in years, N_D and N_P represent the number of atoms of daughter and parent, respectively, and λ is the decay constant in units of 1/year. The natural logarithm function, *ln,* is available on most scientific calculators.

As an example, if N_D = 4000 atoms, N_P = 5000 atoms, and the decay constant for the parent isotope is $\lambda = 1.40 \times 10^{-8}$/ year, then:

$$t = ln(4000 atoms/5000 atoms + 1)/(1.40 \times 10^{-8}/year)$$

$$t = ln(0.8000 + 1)/(1.40 \times 10^{-8}/year)$$

$$t = ln(1.8000)/(1.40 \times 10^{-8}/year)$$

$$t = 0.5878/(1.40 \times 10^{-8}/year)$$

$$t = 4.20 \times 10^7 \text{ years or } 42,000,000 \text{ years}$$

Note that in practice two complications arise. First, daughter isotopes are commonly present in minerals when they first form. Second, surface weathering or metamorphism can cause parent or daughter elements to be gained or lost from minerals within a rock. Both types of complications are easily handled by modern methods, but we will not discuss them here.

Geologic Time Scale

The **geologic time scale** (Fig. 13.7) subdivides Earth's history into unequal intervals based on distinctive fossil assemblages and globally important geologic features, such as widespread unconformities. The scale was developed first as a *relative time scale,* using the principles discussed previously, but numerical ages are now known for all parts of the time scale. The largest subdivisions of geologic time are **eons;** eons are divided into **eras,** eras into **periods,** and periods into **epochs.** The Phanerozoic Eon has many subdivisions because well-preserved shelly fossils are present throughout and are found worldwide. That is not true for the time preceding the Phanerozoic, when organisms were simple life forms without easily preserved shells. Instead, subdivisions of that time, called the **Precambrian** (Fig. 13.7), are based on major geologic features or events and on numerical ages.

Geologic Time Scale

Eon	Era	Period		Epoch	Age (millions of years ago)	Event (Problem 4)
Phanerozoic	Cenozoic (Cz)	Quaternary (Q)		Recent or Holocene		
					0.01	
				Pleistocene		
					1.8	
		Tertiary (T)	Neogene (N)	Pliocene		
					5.3	
				Miocene		
					24	
			Paleogene (Pɛ)	Oligocene		
					34	
				Eocene		
					55	
				Paleocene		
					65	
	Mesozoic (Mz)	Cretaceous (K)				
					144	
		Jurassic (J)				
					206	
		Triassic (Ŧ)				
					248	
	Paleozoic (Pz)	Permian (P)				
					290	
		Carboniferous (C)	Pennsylvanian (ℙ)			
					323	
			Mississippian (M)			
					354	
		Devonian (D)				
					417	
		Silurian (S)				
					443	
		Ordovician (O)				
					490	
		Cambrian (Ꞓ)				
					543	
Precambrian	Proterozoic					
					2500	
	Archean					
					3800	
	Hadean					
					4500	

FIGURE 13.7

The geologic time scale, with symbol abbreviations in parentheses.
Adapted from web page of the Museum of Paleontology of the University of California at Berkeley.

Hands-On Applications

Rocks are the pages in the Earth's history book. Unfortunately, there is no single place on Earth where the whole book is intact. Pages, or even whole chapters, have been locally removed by erosion or bent or torn apart by deformation. The geologist's job is to read the book the way it is, and that requires sorting out the time relations using sound scientific methods. In this lab, you first will learn how basic principles are used to establish the order in which rocks formed or geologic events occurred. Additional problems illustrate how numerical geologic ages are determined.

Objectives

If you are assigned all the problems, you should be able to:

1. Determine relative ages and the sequence of geologic events as illustrated in a geologic cross section.

2. Use daughter-parent ratios, half-life, and the decay constant to determine the numerical age of a rock or mineral graphically and mathematically.

Problems

1. Figures 13.8 and 13.9 are geologic cross sections. Determine the sequence of events that led to the present situation, and list them in order, from oldest to youngest, in the blanks provided. Use the letters on the illustrations to specify rock units or events. Identify the rock types represented by each letter from the symbols given in Figure 13.10; identify other events by name (for example, "folding" or "uplift and erosion"). The number of blanks equals the number of events required.

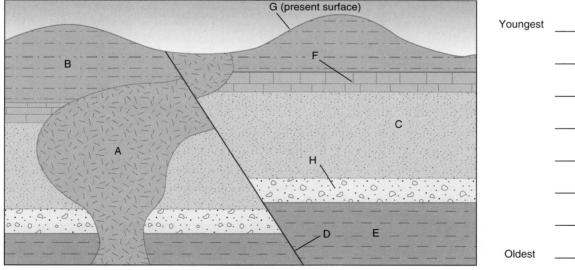

Youngest _____

_____

_____

_____

_____

_____

Oldest _____

FIGURE 13.8

For Problem 1, determine the relative ages of the lettered geologic features illustrated in this cross section, and list them chronologically in the spaces provided. D (dark line) is a fault, and G is the present surface.

Youngest _____

_____

_____

_____

_____

_____

_____

_____

_____

_____

_____

Oldest _____

FIGURE 13.9

For Problem 1, determine the relative ages of the geologic features illustrated in this cross section, and list them chronologically in the spaces provided. Letters indicate rock units; *other events are not labeled.* Also see Problem 6.

Sedimentary rocks

Conglomerate	Breccia	Sandstone	Sandstone
Siltstone	Shale	Limestone	Dolostone

Igneous rocks

Granite	Gabbro	Rhyolite	Basalt

Metamorphic rocks

Slate	Schist	Gneiss	Hornfels adjacent to intrusive contact (cross hatches)

Plutonic rock

FIGURE 13.10

Symbols commonly (but not universally) used to show different kinds of rocks.

2. The purpose of this problem is to illustrate how numerical ages based on radioactivity can be determined graphically.

    a. Complete columns A and B in the table that follows. For example, after one half-life, the parent fraction is 0.5 and the daughter fraction is 0.5; after two half-lives, the parent fraction is 0.25 and the daughter fraction is 0.75. Remember, the sum of the two fractions must equal 1.0.

    b. Next, complete column C in the table by dividing values in Column B by corresponding values in Column A. For example, for one half-life elapsed, (Col. B)/(Col. A) = 0.5/0.5 = 1.

Half-Lives Elapsed	A. Parent Fraction	B. Daughter Fraction	C. Daughter/ Parent Ratio
0			
1	0.5	0.5	1
2	0.25	0.75	
3			
4			
5			

    c. Using Figure 13.11 and the data in the preceding table, construct a graph in which the veritcal axis is "Daughter-Parent Ratio" and the horizontal axis is "Half-Lives Elapsed." The point representing one half-life is already plotted. Plot the rest, and draw a *smooth* curve connecting the data points; that is, do not connect the points with straight-line segments, but estimate the curvature between points as best as you can so that the entire curve bends smoothly.

    d. For samples 1–3 in the following table, first calculate and record the daughter-parent ratio, N_D/N_P. Then, using your graph in Figure 13.11, determine the number of half-lives that have elapsed for each sample and write your answer in the "Half-Lives Elapsed" column.

Sample Number	Atoms of Parent N_P	Atoms of Daughter N_D	N_D/N_P	Half-Lives Elapsed	Age in Years
1	2135	3203			
2	4326	10,815			
3	731	14,620			

    e. If the half-life is 8200 years, calculate the age in years of the samples in the preceding table and write your answer in the "Age in Years" column. Show your work.

3. The purpose of this problem is to illustrate how numerical ages based on radioactivity can be determined mathematically. Show all your calculations.

    a. If the decay constant for radioactive decay of ^{40}K is $\lambda = 5.543 \times 10^{-10}$/year, use Equation 13.1 on p. 251 to calculate the ages of the samples in the following table, and write them in the "Age in Years" column.

Sample Number	Atoms of Parent N_P	Atoms of Daughter N_D	N_D/N_P	Half-Lives Elapsed	Age in Years
1	6439	2303			
2	4395	1303			
3	8763	1893			

Problem 3 is continued on next page.

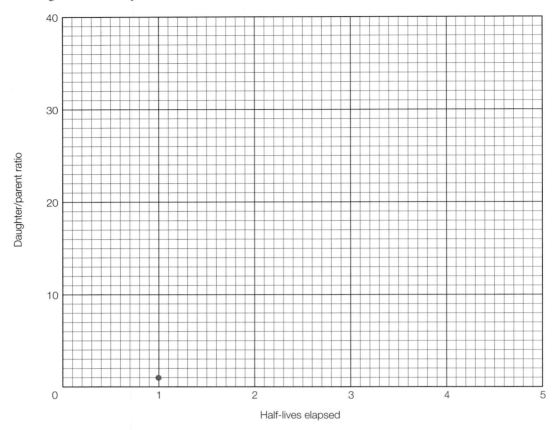

FIGURE 13.11
Graph for Problem 2c.

b. The half-life, $t_{1/2}$, of ^{40}K can be determined from the decay constant using the following relation:

$$t_{1/2} = 0.693/\lambda \ \ (see\ footnote\ 1)$$

What is the half-life of ^{40}K? _____

c. Using the half-life calculated in Problem 3b, complete the column labeled "Half-Lives Elapsed" in the previous table.

4. Go to *www.kaibab.org/geology/gc_layer.htm* (or link to it through *www.mhhe.com/jones6e*—see Preface) and scroll down to the schematic geological section at the bottom of the page. Concentrate on the Precambrian part of the section, that is, everything below the Tapeats Sandstone, and answer the following:

a. Is the Zoroaster Granite older or younger than the Vishnu Schist?

The Bass Limestone?

b. What is the oldest sedimentary rock?

¹The equation $t_{1/2} = 0.693/\lambda$ is derived from the equation $t = \ln\ (N_D/N_P + 1)/\lambda$ as follows: If $t = t_{1/2}$, then $N_D/N_P = 1$, $\ln(N_D/N_P + 1) = \ln 2 = 0.693$, and $\ln(N_D/N_P + 1)/\lambda = 0.693/l = t_{1/2}$

c. What name is applied to the surface separating the Vishnu Schist and the Bass Limestone (be specific)?

What surface separates the Bass Limestone and the Tapeats Sandstone?

d. Knowing that the oldest Precambrian sedimentary rock in the Grand Canyon is about 1,250 m.y. (million years) old and the youngest about 825 m.y., use maximum thicknesses given for the Precambrian sedimentary rocks (that is, exclude the Cardenas Lavas) and calculate the "average" rate of accumulation in feet/m.y.

Evaluate the significance of this number; that is, why might this number be meaningful or meaningless?

Next go to The University of California, Berkeley Museum of Paleontology at *www.ucmp.berkeley.edu/index.php* and under *Online exhibits* click on *Tour of geologic time (www.ucmp.berkeley.edu/exhibits/geologictime.php)*. Use this site to determine when the things in the following list occurred. To do this, click on *Precambrian, Paleozoic Era, Mesozoic Era,* and *Cenozoic Era*. For the three eras, you will have to further click on the *Stratigraphy* buttons to find the information needed. Write the answers below <u>and,</u> to get an overall sense of history, in the right-hand column of the Geologic Time Scale
(Fig. 13.7).

Oldest known rocks?

Transition to atmosphere with oxygen?

Start of breakup of Pangaea?

Oldest known fossils?

First animals?

Age of dinosaurs?

First:
  Birds?
  Reptiles?
  Mammals?
  Insects?
  Fish?
  Flowering plants?

Largest mass extinction?

An <u>excellent</u> interactive site that will help you understand how isotopes are used to determine numerical ages is at *www.sciencecourseware.org/VirtualDating/*. Give it a try.

FIGURE 13.12

For Problem 5, determine the sequence of events illustrated in this cross section. Identify rock units from their symbols, and list all the events that led to the present situation.

In Greater Depth

5.   Using Figure 13.12, identify rock units from their symbols, and identify and list in sequence *all* the events that led to the present situation. On the cross section, write number 1 on the oldest rock or event, number 17 on the youngest, and intermediate numbers on rocks or events in between. If you cannot determine which of two or more rocks/events is the older, explain why not.

6.   Refer to Figure 13.9.

   a.   The daughter-parent isotope ratios (N_D/N_P) for rocks F and C are 16.0 and 9.2, respectively. Assume that the half-life of the parent is 6.5×10^7 years, and use the curve you derived for Problem 2 to determine the numerical ages of each of these rocks.

   b.   During what geologic periods were these rocks intruded (see Figure 13.7)?

   c.   Assume, for illustrative purposes, the highly unlikely scenario in which each of the lettered rock units A through I represents a geologic period of the Phanerozoic. What period (or periods, if more than one answer is possible) does each represent?

   d.   During what geologic period or periods did folding and erosion occur?

Part VI
Internal Processes

Folded layers in the banded iron formation, Jasper Knob, Ishpeming, Michigan (×2)

CHAPTER 14

Structural Geology

Materials Needed

- Pencil and eraser
- Protractor
- Metric ruler
- Silly Putty® (provided by instructor)
- Scissors
- Removable tape (optional)

Introduction

Stresses within the Earth bend and break rocks, producing folds, faults, and other geologic structures. These structures are three-dimensional, and they are commonly too large or too poorly exposed to be easily recognized from a single vantage point on the ground. For these reasons, basic geologic data must be collected in the field and plotted on a map to understand the structure fully. This chapter will help you visualize geologic structures in three dimensions so that you can understand them when seen in two dimensions on a geologic map, aerial photograph, or on the Earth's surface. You will learn how the orientations of planar features, such as sedimentary bedding, are described and used to portray the geologic structure on a geologic map. Vertical cross sections of geologic maps illustrate what you expect to see below the surface, and block diagrams combine maps and cross sections to provide a three-dimensional view. A clear understanding of subsurface structures is vital for an understanding of groundwater resources, pollution migration, oil reserves, ore deposits, and the stability of road

cuts, tunnels, bridge footings, and building foundations.

Structural geology is the study of the form, arrangement, and internal structure of rocks. It is especially concerned with understanding how rocks are deformed. Evidence for deformation ranges from rocks shattered during earthquakes, to rocks long ago tilted, folded, and shoved over one another, to the foliation seen in slates, schists, and gneisses. The structural geologist puts together all such evidence to unravel the remarkable complex interactions of the Earth's tectonic plates.

Deformation is caused by **stress,** a force acting on an area of a body that tends to change its size or shape. Stress is a force per unit area, so the magnitude of stress depends not only on the magnitude of the force, but also on the area over which the force is applied. Stress will be greater if a given force is applied to a small, rather than large, area; a person wearing high-heeled shoes is more likely to punch a hole in a rotten floorboard than a person wearing tennis shoes. *Compressive stress,* or **compression,** causes shortening (Fig. 14.1A), and is an important type of stress at convergent tectonic plate boundaries. *Tensional stress,* or **tension,** the principal stress at di-

vergent boundaries, causes stretching (Fig. 14.1B). **Shear stress** causes one side of a body to slide past the other (Fig. 14.1C), and is the principal type of stress along transform-fault plate boundaries.

Strain, the physical deformation that occurs in response to stress, is a change in size or shape. Strain may be temporary or permanent. **Elastic strain** is temporary; if the stress is removed, the object snaps back to its original shape. A stretched rubber band and a bouncing ball both undergo elastic strain. **Plastic strain** is permanent; if the stress is removed, the original size or shape is not restored. If you bend a piece of soft wire, the wire undergoes plastic strain; the bend is permanent. If it is bent back and forth at the same place, it eventually breaks or **ruptures.**

A rock undergoing elastic strain can suddenly break: this is **brittle deformation.** A dry stick of wood also behaves this way: it bends only so far before it suddenly breaks (ruptures), and each piece elastically recovers its original straight shape. A rock that undergoes much plastic strain without rupturing is **ductile.** The aforementioned bent wire is ductile.

Pressure, temperature, and rate of deformation determine brittle versus ductile

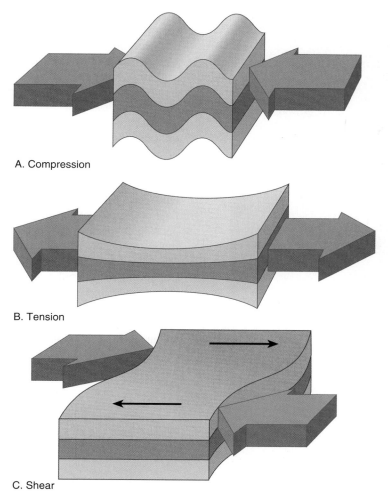

A. Compression

B. Tension

C. Shear

FIGURE 14.1

The three principal types of stress: A. compression; B. tension; C. shear.

Table 14.1		
Conditions Required for Brittle versus Ductile Deformation of Rocks		
Surrounding Conditions	*Brittle*	*Ductile*
Pressure	Low	High
Temperature	Low	High
Stress rate	Rapid	Slow
Stress magnitude	High	Low to high

bend into folds. Eventually, erosion of overlying rock material and gradual uplift bring the deformed rocks to the surface where we can view them.

Recognizing and Describing Deformed Rocks

A geologist must be able to visualize rock structures in three dimensions. In some parts of the world, rocks are so well exposed at the surface that it is easy to see and understand their structure. In most areas, however, the rocks are buried beneath soil or young, unconsolidated sediment, and are exposed only in a few **outcrops** located far apart. In such places, it is difficult to understand the three-dimensional aspects without carefully examining the rocks, measuring their orientation, and plotting these data on a map.

What to Look for: Stratification, Contacts, and Fault and Joint Surfaces

How can you tell if sedimentary rocks have been deformed? It generally is easy—just look at the orientation of the strata. The original sediments were deposited as continuous horizontal or nearly horizontal **strata** (layers). If the strata are still horizontal and not broken, then they have not been deformed, although they may have undergone gentle uplift or subsidence. If strata are broken or not horizontal, then they have been deformed. Deformation takes many forms, from slight tilting of strata to complex folding, as described in a later section.

Contacts are surfaces separating adjacent bodies of rock of different types or ages. Understanding the nature of the contact is important, because the contact records a change, an event. There are several kinds of contacts.

Depositional contacts separate older bodies of rock from younger sedimentary rocks deposited upon them. In a sequence of sedimentary rocks deposited continuously, or **conformably,** in a basin of deposition, the contacts between adjacent layers are planar **bedding planes** (see Fig. 4.3); they are parallel to—and, in fact, define—the bedding or stratification. Some depositional contacts are ancient

behavior (Table 14.1). Cold rocks at the Earth's surface that experience a rapidly applied stress (like a hammer blow) tend to experience brittle fracture. Ductile deformation is more likely under the high temperature and pressures found deep underground, especially when stresses are

applied slowly. The folded rocks visible in many mountain ranges were deformed well below the surface by forces acting over long periods of time (millions of years). The combination of depth and time enables deep rocks that would be brittle at the surface to behave plastically and gradually erosion surfaces. These **unconformable** contacts can be uneven rather than planar, and can juxtapose very different rocks with very different orientations (see *Unconformities* in Chapter 13).

An *intrusive contact* separates intrusive igneous rock and the rock that it intruded; intrusive contacts are either **concordant** (parallel to layers in the intruded rock) or **discordant** (cut across layers in the intruded rock).

Rocks separated by a fault are in *fault contact* (Fig. 14.2). In the simplest case, the fault is exposed as a flat surface, and its orientation is apparent. However, faults commonly are expressed as zones along which rocks are brecciated (broken into angular fragments), or even pulverized, so that fault orientation may be difficult to determine. Even more commonly, faults, like other types of contacts, are buried beneath soil or sediment. In such places, their orientation,

FIGURE 14.2

The steeply inclined surface in the center of the photograph is a fault contact. It cuts the sandstone and mudstone layers of the Haymond Formation (Pennsylvanian) near Marathon, Texas. Width of area shown is about 1.5 m.

A. Outcrop

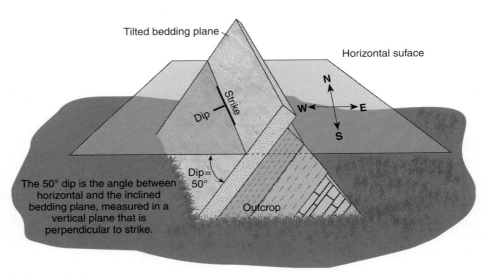

Tilted bedding plane

Horizontal suface

Strike

Dip

N

W   E

S

The 50° dip is the angle between horizontal and the inclined bedding plane, measured in a vertical plane that is perpendicular to strike.

Dip= 50°

Outcrop

B. Perspective view

30°

Compass

90° W   E 90°

0°
N
30°
SW
Quad.
S
180°

50°

C. Map (overhead) view

FIGURE 14.3

A. Outcrop of Mississippian Bluefield Formation near Covington, Virginia. The sedimentary layers are inclined to the left, or southwest. B. Simplified diagram showing the inclined layers protruding through a horizontal plane. The intersection of the inclined layers with the horizontal plane creates a line that defines the strike of the bed. This line trends 30° west of the north arrow. The beds dip 50° below the horizontal plane. C. When viewed from above, as on a map, the strike and dip of the bed are shown with a T-like symbol. The top of the T parallels the strike, and the leg of the T points toward, and parallels, the dip direction. The dip angle is 50°. Comparison of the strike and dip symbol with the compass directions shows why the attitude of this bed is written as N30°W, 50°SW.

and even their presence, must be inferred from data gathered from many outcrops and plotted on a map.

Describing Surface Orientation

You will see on geologic maps curious little T's called **strike and dip** symbols (Fig. 14.3). These are used to describe the orientation of bedding planes and other planar features so that the overall geometry of tilted and folded layers can be described and visualized on maps. The long top part of the T parallels the **strike,** which represents the intersection of a tilted bedding plane with a *horizontal* plane (Fig. 14.3B). It gives the trend of that plane. For example, if you hold a credit card at a 45° angle to a table and draw a line along where it touches the table, you get the strike of the plane defined by your credit card. Hold the card against the strike line and notice how the strike indicates which way a plane trends, but not which way it tilts. The **dip direction** of a plane is shown by a short

line drawn perpendicular to the strike and pointing downhill (Fig. 14.3B). Draw a short line perpendicular to your strike, away from the credit card, and you have the dip direction. Write, for example, 45° (the **dip angle**) at the end of your dip line and you uniquely define the orientation of your credit card. You can easily visualize the orientation of a plane described by a strike and dip symbol by holding your hand parallel to the strike and tilted down toward the dip. Try it on the perspective drawings later in the chapter.

A geologist measures the orientation, or **attitude,** of a tilted bedding plane by holding a compass in a horizontal position, with a straight side against the bedding plane (Fig. 14.3C). While the compass arrow points north, the compass body points in the direction of strike, a certain number of degrees east or west of north. Figure 14.3C shows a strike line pointing 30° west of north, which is written as N30°W. Strike measurements are commonly given with the number of degrees less than 90. Thus, N95°E

becomes N85°W, and N90°E is written E-W.

The dip angle, or **angle of inclination,** is measured from the horizontal plane down to the bedding plane (Fig. 14.3B). The measuring device must be held in a vertical plane that is perpendicular to the strike in order to be accurate. Because a dip can point either direction from a strike line, it is necessary when recording dips to specify the dip direction. Since the dip is always perpendicular to strike, it is necessary only to indicate the compass quadrant toward which the surface dips (that is, NW, NE, SE or SW). For example, with a strike of N30°W, the dip direction could either be NE or SW (Fig. 14.3C). In the example used in Figure 14.3B, the complete strike and dip of the bedding plane is written N30°W, 50°SW.

Geologists also measure the attitudes of fault planes, fracture surfaces, and other geologic features. Figure 14.4 shows special symbols for vertical, horizontal, and overturned sedimentary layers, plus symbols for folds and faults. The arrows used in

FIGURE 14.4
Geologic map symbols.
Source: American Geological Institute and others.

the fold and fault symbols point in the direction of dip.

Geologic Maps, Cross Sections, and Block Diagrams

Geologic information is compiled and displayed in three common ways, shown in Figure 14.5: geologic maps, geologic cross sections, and block diagrams.

Geologic maps show the distribution of rock units at or very near the surface of the Earth (Fig. 14.5A). They also portray the structure of the rock units by means of strike and dip symbols. An accompanying legend, or **explanation,** lists and briefly describes the rock units in order of age; major periods of erosion are also listed. Topography may be shown with an overlay

of contour lines. The next chapter focuses on geologic maps.

Geologic cross sections are vertical slices that show how the rocks would appear if they could be viewed on a cliff face (Fig. 14.5B). They are topographic profiles with the area below the surface filled in with representations of rocks. Because such cut-away views do not usually present themselves in nature, they are interpretations based on information obtained on the surface or from wells or geophysical data.

Block diagrams combine maps and cross sections to show the three-dimensional aspects (Fig. 14.5C). The three-dimensional blocks are drawn as two-dimensional perspective drawings, with the map view shown on the top and cross sections on the ends and sides. Most block diagrams have a flat upper surface and do not attempt to

show topography, but some, drawn by clever people, manage to illustrate topography as well as geology.

Folds

A **fold** is a bend in a layer of rock due to ductile (plastic) strain. Upward bends, or upfolds, with older rocks in the center, are called **anticlines;** downward bends, downfolds, with younger rocks in the center, are called **synclines** (Fig. 14.6). The sides or flanks of a fold are the **limbs.** The limbs join at the **hinge line** of the fold, which is the line of maximum curvature (Fig. 14.6). The **axial surface** (or **axial plane** if it's not curved) of a stack of folded layers passes through the hinge lines and most nearly divides the fold into two equal parts (Fig. 14.6). An **upright fold** has a vertical

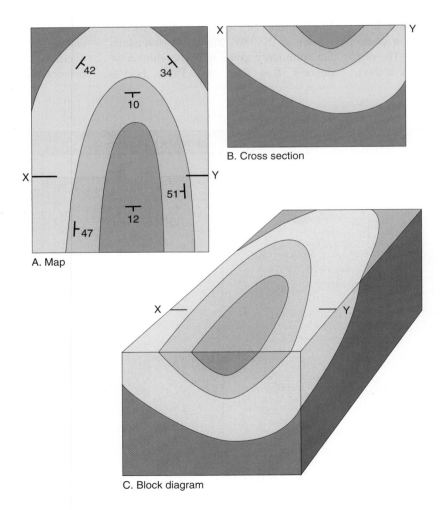

A. Map

B. Cross section

C. Block diagram

FIGURE 14.5

Three ways in which geologic information is illustrated: A. A geologic map shows rock units and geologic structure as they appear on the surface. You can use your hands to visualize the orientation of the bedding described by the strike and dip symbols. B. A geologic cross section of the map from X to Y is like a vertical slice into the Earth. C. Block diagrams combine map and cross-section views in a three-dimensional perspective drawing.

or nearly vertical axial surface; beds on opposite limbs have similar dips, though in opposite directions (Fig. 14.7A). An **inclined fold** has an axial surface that is neither vertical nor horizontal (Fig. 14.7B). An inclined fold is **overturned** if beds on opposite limbs dip in the same direction; beds on the overturned limb are upside-down, as they have been rotated more than 90° (Fig. 14.7C). A **recumbent fold** has an approximately horizontal axial surface (Fig. 14.7D).

A **non-plunging fold** has a horizontal or nearly horizontal hinge line (Fig. 14.8A, B). A **plunging fold** has an inclined hinge line (Fig. 14.8C, D).

Figure 14.9 illustrates folds using block diagrams in which the top surface is horizontal and flat. Note that when non-plunging folds intersect a horizontal

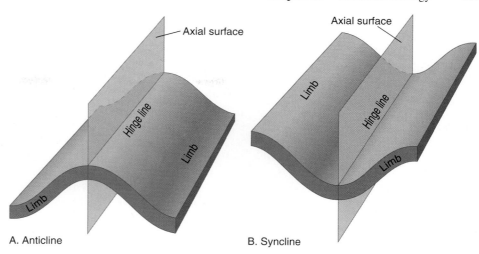

A. Anticline   B. Syncline

FIGURE 14.6

Folds are described in terms of limbs, hinge line, and axial surface, shown here for an anticline (A) and a syncline (B).

A. Upright   B. Inclined

C. Overturned   D. Recumbent

FIGURE 14.7

The axial surface of a fold can be: A. Vertical in upright folds; B. inclined in inclined folds; C. inclined so much that opposite limbs dip in the same direction in overturned folds; D. horizontal in recumbent folds.

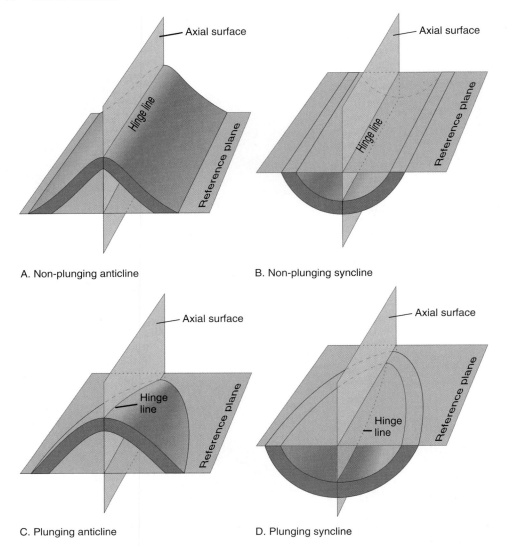

A. Non-plunging anticline

B. Non-plunging syncline

C. Plunging anticline

D. Plunging syncline

FIGURE 14.8

Non-plunging folds have horizontal hinge lines, as in a non-plunging anticline (A) and a non-plunging syncline (B). Plunging folds have inclined hinge lines, as in a plunging anticline (C) and a plunging syncline (D). Axial surfaces are vertical, and reference plane is horizontal.

surface, such as that approximated by the surface of the Earth, the contacts between adjacent beds are straight lines (Fig. 14.9A, B). When plunging folds intersect a horizontal surface, the contacts between adjacent beds are curved lines. In a plunging anticline, the contacts bend so that they "point" in the direction of plunge, whereas in a plunging syncline, they "point" in the direction opposite the plunge (Fig. 14.9C, D).

Using the Principle of Superposition (Chapter 13), you can see that in an eroded syncline, the *youngest* beds appearing on the surface are in the center of the syncline (Fig. 14.9B, D). In an eroded anticline, the *oldest* (originally deepest) beds are in the center (Fig. 14.9A, C).

Domes and **basins** have cross sections like anticlines and synclines, respectively, but are approximately circular or oval in map view (Fig. 14.10). Beds in domes dip outward in all directions, and like anticlines, the oldest rocks are found in the center. Similarly, the youngest rocks in basins are in the center, just as they are in synclines, and beds dip inward. Domes and basins may be quite large, with diameters of 100 km or more.

Faults

Faults are breaks or fractures along which movement has occurred. The rocks on one side of a fault have moved relative to those

on the other side. Relative movement may be up and down, sideways, or a combination of the two. The amount of movement ranges from centimeters to kilometers.

Faults with inclined fault planes have **footwalls** and **hanging walls** (Fig. 14.11). The **footwall** is under the fault plane: if you dug a mine along the fault plane, you would stand on the footwall. The **hanging wall** is above the fault plane: in a mine, it would hang above your head.

Faults are classified according to their relative motion. When the hanging wall has dropped relative to the footwall, it is a **normal fault** (Fig. 14.11A). When the hanging wall has been pushed up relative to the footwall, it is a **reverse fault** if the fault

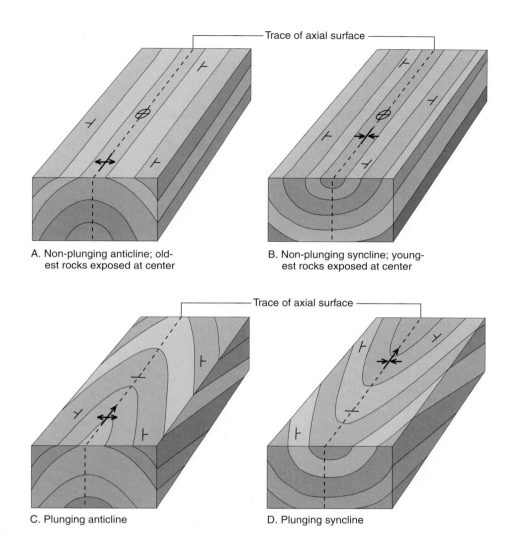

Trace of axial surface

A. Non-plunging anticline; oldest rocks exposed at center

B. Non-plunging syncline; youngest rocks exposed at center

Trace of axial surface

C. Plunging anticline

D. Plunging syncline

FIGURE 14.9

The top surface of these block diagrams, which is analogous to the Earth's surface, shows that older beds are exposed in the centers of eroded anticlines and younger beds are exposed in the centers of eroded synclines. Contact lines on the top surface are straight if folds do not plunge, but bend if they do. In plunging anticlines, the contact lines bend and point in the direction of plunge; in plunging synclines, they bend and point away from the direction of plunge. In C and D, both folds plunge into the page. The symbols on the top surface show the strike and direction of dip at various locations and the axial traces and type of folds. Note how these symbols help you visualize structure from map view alone.

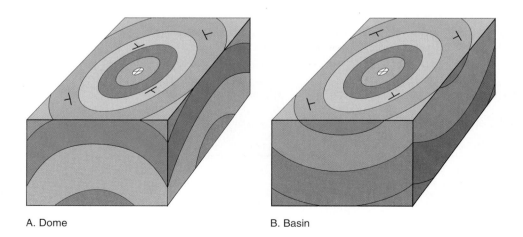

A. Dome

B. Basin

FIGURE 14.10

These block diagrams show that older beds appear in the center of an eroded dome, while younger beds appear in the center of an eroded basin. The symbols on the top surfaces show the strike and direction of dip at various locations.

plane is steeply dipping (more than 45°; Fig. 14.11B) or a **thrust fault** if the fault plane has shallow dip (less than 45°; Fig. 14.11C). **Lateral** or **strike-slip faults** have almost vertical fault planes and show off-sets in a horizontal plane (Fig. 14.11D). *Right-lateral strike-slip faults* are those in which the block on the opposite side of the fault has moved relatively to the right. Thus, if you stood on one side of a right-lateral strike-slip fault, and looked across the fault, the rocks on the other side would be displaced to the right, as in Figure 14.11D. In *left-lateral strike-slip faults,* the block on the opposite side of the fault has moved relatively to the left. Whereas the other faults are most obvious in vertical outcrops, offsets along strike-slip faults are best seen from the air.

Clues to the Past

Folds and faults record the history of geologic stresses in a given area. Such stresses may result, for example, from continental blocks pulling apart, sliding by one another, or colliding. Table 14.2 summarizes the types of stress and strain that are recorded by folds and the different types of faults. Assessing the type of stress and the directions from which it was applied helps us to unravel the long and complicated series of events that have shaped and reshaped large areas of especially eastern, southern, and western North America.

A. Normal fault

B. Reverse fault

C. Thrust fault

D. Right-lateral strike-slip fault

FIGURE 14.11

Four types of faults are illustrated in these block diagrams. If the fault plane is inclined, the side above the fault is the hanging wall, and the side below is the footwall. Displaced marker beds show the direction of movement. Note that in a map view, shown by the top surfaces of A, B, and C, the faults dip under the hanging wall, and in C the barbs or "teeth" are on the hanging wall.

Table 14.2
Stress/Strain Conditions of Common Geologic Structures

Strain	Stress		
	Compression	Tension	Shear
Brittle strain or rupture	Reverse fault or thrust fault	Normal fault	Strike-slip fault
Ductile or plastic strain	Folds (anticlines and synclines)	Not discussed in text	Not discussed in text

Hands-On Applications

If geologists are to understand a dynamic Earth, they must understand the nature of the stresses within the Earth. One way to learn about stress is to study the strain, or deformation, that it has produced. This type of study is based on descriptive and interpretive analysis of folds, faults, and other deformational features. Using drawings and three-dimensional models, this lab focuses mainly on the geometric aspects of deformed rocks. You will learn how observations and measurements made at the surface are used to predict the structure of rocks in the subsurface.

Objectives

If you are assigned all the problems, you should be able to:

1. Distinguish between elastic and plastic strain, and list the conditions that might favor one over the other.

2. Recognize structures formed by compression, tension, and shear.

3. Identify conformable and unconformable depositional contacts, intrusive contacts, and fault contacts on a geologic map, cross section, or block diagram.

4. Define dip and strike, plot them on a map, and determine their general orientation on a geologic map or top surface of a block diagram.

5. Define, sketch, and recognize—on a geologic map or a block diagram—a dome or basin and plunging and non-plunging anticlines and synclines.

6. Determine the direction of plunge of plunging anticlines and synclines on a geologic map or block diagram.

7. Define, sketch, and recognize—on a cross section or block diagram—normal, reverse, thrust, and strike-slip faults.

8. Distinguish the hanging wall and footwall of a normal, reverse, and thrust fault on a cross section, geologic map, or block diagram.

Problems

1. To demonstrate how the same material can respond to stress (deform) differently under different conditions, try the following with Silly Putty™. First, roll it into a ball and see if it bounces. Next, roll it into a thick cigar shape and slowly stretch it out. Finally, roll it into a thick cigar shape again and pull it apart as fast as you can. In each case, the Silly Putty behaves differently.

   a. Name the type of behavior exhibited in the three cases, using the terminology given in the introduction to this lab.

   b. How did the conditions under which deformation occurred differ in the three cases?

Your instructor may want you to do either Problem 2 (cut-out blocks) or Problems 3 and 4 (perspective drawings).

2. The following problems use the cut-out block diagrams A, B, and C, which are found at the end of the lab manual. The blocks should be cut out before beginning the problem in which they are used. When constructing the box, be very careful to make the folds exactly along the lines, or the problems may not work out as they should. It will be easier to draw on the boxes if you do not tape the sides together until after you have completed the problem. Even then, it is recommended that you use removable tape.

   a. Refer to Block A with Side 1 up. When Side 1 is up, Block A is similar to Figure 14.9A, and the top surface represents the Earth's surface.

      (1) Fill in the blank faces on the sides of the block.

      (2) Measure the dip with a protractor at several places on the front surface (note that the dip changes below the surface and that the angle you should measure is right at the top surface) and put strike and dip symbols at the places indicated with dots on the top (map) surface.

(3)   Illustrate the position of the axial surface of the fold by sketching its trace on the top and ends of the block (the **trace** is the imaginary line formed where the axial surface intersects other surfaces).

(4)   When Side 1 is up, what is the age of the beds in the center, or core, of the fold relative to the other beds?

(5)   What kind of fold is this (anticline or syncline; upright, inclined, overturned, or recumbent; plunging or non-plunging)?

b.   Refer to Block A with Side 2 up. This is a new situation in which Side 2 now represents the Earth's surface.

(1)   Put strike and dip symbols at the places indicated with dots on the top (map) surface.

(2)   When Side 2 is up, what is the age of the beds in the center, or core, of the fold relative to the other beds?

(3)   What kind of fold is this?

(4)   Sketch the trace of the axial surface (axial trace) on the top of the block using one of the fold symbols in Figure 14.4.

c.   Refer to Block B with Side 1 up. When Side 1 is up, Block B is similar to Figure 14.9C, and the top surface represents the Earth's surface.

(1)   Fill in the blank faces on the side of the block.

(2)   Put strike and dip symbols at the places indicated with dots on the top (map) surface.

(3)   Illustrate the position of the axial surface of the fold by sketching its trace on the top and ends of the block.

(4)   When Side 1 is up, number the beds from oldest (number 1) to youngest. Where are the oldest beds as seen on the top (map) surface?

(5)   What kind of fold is this?

d.   Refer to Block B with Side 2 up. This is a new situation in which Side 2 now represents the Earth's surface.

(1)   Put strike and dip symbols at the places indicated with dots on the top (map) surface.

(2)   When Side 2 is up, what is the age of the beds in the center, or core, of the fold relative to the other beds?

(3)   What kind of fold is this?

(4)   Sketch the trace of the axial surface (axial trace) on the top of the block using one of the fold symbols in Figure 14.4.

e.   Refer to Block C, with Side 1 up.

(1)   Complete the sides of the block.

(2)   Put strike and dip symbols of beds at the dots on the top (Side 1) surface.

(3)   Put a dip symbol (Fig. 14.4) on the fault (dark line).

(4)   Indicate the relative movement of the fault by placing arrows on opposite sides of the fault on the side faces.

FIGURE 14.12
Block diagram to be completed for Problem 3.

   (5)   Label the hanging wall and footwall on the top and side faces.

   (6)   With Side 1 up, number the beds in order of age, with 1 the oldest. Are the beds at the surface of the upthrown block generally older or younger than those at the surface of the downthrown block? Why?

   (7)   What kind of fault is this?

3.   Figure 14.12 is a block diagram showing a fault and a number of folds. The front face of this diagram is perpendicular to the strike of these features, so you can easily measure the dip of the fault and bedding planes where indicated. North is into the page, west and east are marked.

   a.   Use a protractor to measure the dip angles along a single fold at points 1 through 5 on the front face of the diagram. Horizontal lines indicate the precise location for each measurement.

   1.   2.   3.   4.   5.

   b.   A geologist would take strike and dip measurements at convenient outcrops on the surface. Take dip measurements along the front of the block diagram, as close to the surface as possible. Put the appropriate strike and dip symbols in the six white ovals along the top of the diagram, and write each dip measurement in the appropriate spot.

   c.   Draw dashed axial traces for the three folds not cut by a fault. Put them along both the vertical and horizontal surface, and include the appropriate symbol from Figure 14.4.

   d.   What type of fault is illustrated? Be sure the correct map symbol is used to indicate dip. Put arrows on the vertical surface and map symbols on the horizontal surface to indicate the relative sense of motion across the fault.

4.   Partially completed block diagrams are shown in Figure 14.13. Using the information provided for each diagram, fill in the blank faces on the blocks and answer the questions.

   a.   The vertical column on the front face of Figure 14.13A shows the different rock units encountered in an old oil well.

   (1)   Use the available information to help you complete the diagram.

   (2)   What kind of fold is this (anticline, syncline; upright, inclined, overturned, or recumbent; plunging or non-plunging)?

   (3)   Illustrate the position of the axial surface of the fold by sketching its trace on the top and end of the block (the **trace** is the imaginary line formed where the axial surface intersects another surface).

FIGURE 14.13
Block diagrams to be completed for Problem 4.

b. Again, the vertical column on the front face of Figure 14.13B shows the different rock units encountered in an old water well.

(1) Use the available information to complete the front and as much of the side face *as possible*. (Don't go beyond what the observations allow!)

(2) Much of the side face is blank even after using the information provided. Propose a site for another well that would provide the maximum information about that blank side, and fill out the side assuming that two new units were encountered in your well.

(3) What kind of fold is this (anticline, syncline; upright, inclined, overturned, or recumbent; plunging or non-plunging)?

(4) Illustrate the position of the axial surface of the fold by sketching its trace.

c. The vertical column on the front face of Figure 14.13C shows the different rock units found in a recently drilled oil well. A fault is shown on the top face.

(1) Use the available information to help you complete the diagram.

(2) Label any folds on the diagram using the usual terminology.

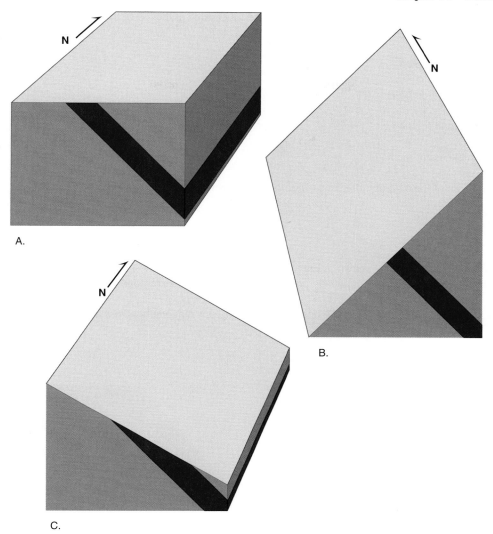

FIGURE 14.14

These block diagrams for Problem 5 show a bed dipping at 45°E, but from slightly different perspectives (note north arrows). A. The upper surface is horizontal, so it intersects the bed at 45°. B. The upper surface slopes at 45°W, so it intersects the bed at 90°. C. The upper surface slopes 30°E, so it intersects the bed at 15°.

(3) Indicate the sense of motion along the fault in Figure 14.13C by putting arrows on the vertical surface. Which is the hanging wall and which the footwall?

(4) What type of fault is indicated in Figure 14.13C (normal, reverse, thrust, strike-slip)?

5. You may have noticed on the various faces of the cut-out models or on the block diagrams of Figure 14.9 that the width of a given layer changes depending on the angle at which the layer intersects the face. This problem is intended to show you why. The three block diagrams in Figure 14.14 show a north-south-striking bed of the same thickness dipping at 45°E. In addition to being drawn from slightly different perspectives, the diagrams also differ because in *A* the upper surface is horizontal and intersects the bed at 45°, in *B* the top surface slopes 45°W so that it intersects the bed at 90°, and in *C* it slopes 30°E so that it intersects the bed at 15°.

a. Complete the upper surface of the diagrams by extending the bed across that surface. The resulting band on the surface represents the outcrop of that bed.

    b.   Measure the following with a ruler (the outcrop width is best measured at the top of the front face of the box, where the front face intersects the upper surface):

        (1)   width of the outcrop on *A;*

        (2)   width of the outcrop on *B;*

        (3)   width of the outcrop on *C;*

        (4)   the true thickness of the bed.

    c.   Why does the width of the outcrop differ in each of the diagrams when the true thickness of the bed is the same?

6.   Partially completed block diagrams are shown in Figure 14.15. Complete each of these by filling in the blank faces. Place general dip and strike symbols on the top surface if not already present. Lowest numbers correspond to oldest rocks.

    a.   In Figure 14.15A, what name is given to the uneven feature that runs diagonally across the map surface?

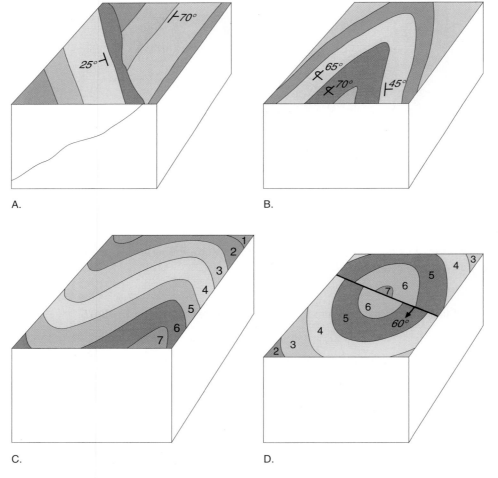

FIGURE 14.15
Block diagrams to be completed for Problem 6. Lowest numbers are oldest units.

b.   What structure is shown in Figure 14.15B?

c.   Describe the structure shown in Figure 14.15C.

d.   What kind of fault is shown in Figure 14.15D, and what kind of structure does the fault cut?

7.   Fill in the blanks to formulate a series of rules to help in the interpretation of geologic maps and cross sections.

a.   The following rules refer to relative ages of sedimentary beds. Choose *older* or *younger* for your answers.

• When anticlines and domes are eroded, the beds exposed in the center are _____ than the beds exposed away from the center.

• When synclines and basins are eroded, the beds exposed in the center are _____ than the beds exposed away from the center.

• On a geologic map, _____ beds dip under (or toward) _____ beds, unless they are overturned. (Hint: Look at the upper surfaces of your cut-out block diagrams or Figure 14.9.)

b.   The following rules refer to beds and contacts between beds, as seen in map view.

• When _____ folds have been eroded, the contacts between beds, as seen on a horizontal surface, are straight and parallel. Choose *plunging* or *non-plunging.*

• When plunging anticlines have been eroded, the contacts between beds, as seen in map view, bend so as to point _____ the direction of plunge. Choose *toward* or *away from.*

• When plunging synclines have been eroded, the contacts between beds, as seen on the surface, bend so as to point _____ the direction of plunge. Choose *toward* or *away from.*

• The outcrop width of a bed _____ as the angle between the surface and the dip of the bed decreases. Choose *increases* or *decreases.*

8.   Many state geological surveys post geologic and other maps on the Web. Go to *www.ohiodnr.com/geosurvey/* (or link to it through *www.mhhe.com/jones6e*—see Preface) and link to *Geology of Ohio,* then *Maps,* and finally *The Old State Bedrock Map.* Because of the scale of the map and the low relief of Ohio, the map is comparable to the top surface of a block diagram. As indicated by the explanation, the rocks are grouped into the geologic time periods in which they were deposited. If you've forgotten the relative ages of geologic periods, consult Figure 13.7.

a.   Where are the youngest units in the state and what is their geologic age?

Where are the oldest units in the state and what is their geologic age?

b.   Which way do the rock units in the eastern half of the state dip and how do you know?

Which way do the rock units in the westernmost part of the state dip and how do you know?

c.   What is the large structural feature centered in the western half of the state and how do you know?

In Greater Depth

9.   The following problems refer to cut-out block diagram D, found at the end of the lab manual.

   a.   Complete the following:

   (1)   Fill in the blank side panels.

   (2)   Put strike and dip symbols at the places indicated with dots on the Side 1 surface.

   (3)   Measure the dip with a protractor on:

   (a)   the small, already filled-in, front panel;

   (b)   the two corner panels that you filled in;

   (c)   the side surface. The small front panel is perpendicular to strike, so the dip measured there is the *true dip;* the other surfaces are not perpendicular to strike, so the dips observed and measured there are not true dips but *apparent dips.*

   (4)   Fill in Side 2.

   (5)   Imagine that the top half of the box (either side up) was eroded away and you could see the new surface exposed. In what direction, in terms of strike or dip, would the outcrops have shifted from their pre-erosion positions?

   b.   Fill in the blanks to formulate additional rules to help in the interpretation of geologic maps and cross sections.

   • The apparent dip of beds seen on vertical surfaces _____ as the angle between the strike of the beds and the strike of the vertical surface decreases (choose *increases* or *decreases*); the true dip can be seen only on vertical surfaces that are _____ to the strike of the beds (choose *parallel* or *perpendicular*).

   • As erosion continuously lowers the level of the land surface, a dipping bed or fault, as seen in map view, migrates _____ the direction of dip (choose *in* or *away from*).

10.   The three cross sections in Figure 14.16 show three outcrops of the same graywacke (sandstone) unit, all of which have beds that dip about 60° to the west. The surrounding rocks, probably shale, are easily eroded and are covered with soil.

   a.   Formulate three hypotheses to explain the relations among the outcrops by completing the cross sections in three different ways. Use dashed lines to indicate contacts above the cross section that have been eroded away.

   b.   What kind of information might the three outcrops contain that would enable you to distinguish among your multiple hypotheses? (Hint: see *Sedimentary Structures* in Chapter 4.)

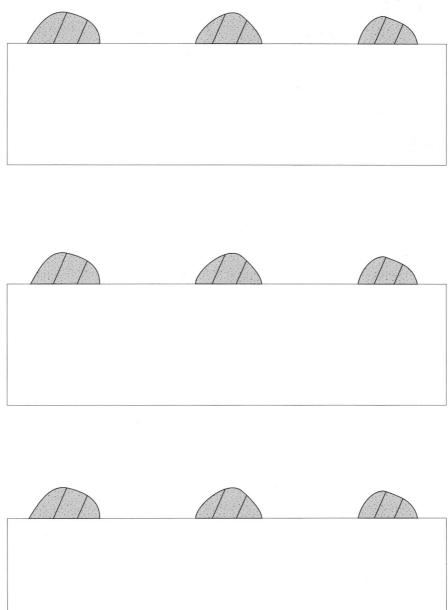

FIGURE 14.16
Cross sections for Problem 10.

CHAPTER 15

Geologic Maps

Materials Needed

- Pencil and eraser
- Ruler
- Protractor

- Calculator
- Stereoscope (provided by instructor)

Introduction

Geologic maps are a visual expression of the basic field data of geology. A good geologic map shows not only what kinds of rocks are present, but also their ages and structure. Because geologic maps are two-dimensional views of three-dimensional rock sequences, they can appear quite complex. However, simple observations can help you understand what is shown on a geologic map. You will learn: how geologic maps are made; how to recognize whether rock formations are horizontal, inclined, folded, or faulted, even if dip and strike symbols are absent; how to determine the chronologic order of geologic events; and how to visualize the third dimension by drawing a geologic cross section.

A geologic map shows the distribution and structure of rock units on or very near the Earth's surface. An accompanying *explanation* lists the rock units present and explains special symbols. Geologic maps commonly are drawn on contour maps so that geology and topography are easily compared.

The rock units shown on geologic maps are usually *formations,* although in some cases, other types of rock units or time-rock units (rocks formed during a specific time interval) are shown. A **formation** is a mappable body of rock. That is, it has an appearance or composition that distinguishes it from other nearby rocks, plus distinct boundaries **(contacts)** that make it easy to separate from adjacent rock units. Formations have two-word names. The first word normally refers to the place where the formation originally was described. If the formation consists principally of one type of rock, the rock name is used as the second part; if not, the word "formation" is used. Examples: St. Peter Sandstone, Trenton Limestone, Morrison Formation.

Making a Geologic Map

Making a geologic map requires:

1. collecting the basic data in the field;
2. correlating among outcrops;
3. making a geologic cross section of the area to show the three-dimensional relation of one rock unit to another; and
4. preparing an *explanation.*

Collecting the basic data is by far the most important and time-consuming task—and the most fun. You can locate expo-sures, or **outcrops,** of bedrock by studying aerial photographs and by walking the countryside to locate where the rock actually can be seen. In the western United States, rock is exposed over large areas. In more humid areas, like the U.S. Midwest, outcrops are commonly small and widely scattered; natural outcrops commonly are found in stream valleys, but rock is also exposed in roadcuts, mines, quarries, and construction sites. The outcrops are examined in the field and plotted on a topographic map or an aerial photograph. A description is recorded on the map or in a field notebook and includes such information as rock type, formation name, strike and dip of beds, faults, folds, or other structural features, fossils present, and anything else deemed important by the geologist. Figure 15.1A is an example of a map showing outcrops; the area between outcrops is covered with soil.

The next step is correlation among outcrops; this is usually done in the field. Notice in Figure 15.1A that strikes and dips are similar in all outcrops and that the sandstone labeled *Pc,* in particular, occurs in a series of outcrops strung out in the same orientation as the strike. The simplest and most reasonable assumption to make is that the four outcrops of this sandstone belong to the same formation and should be

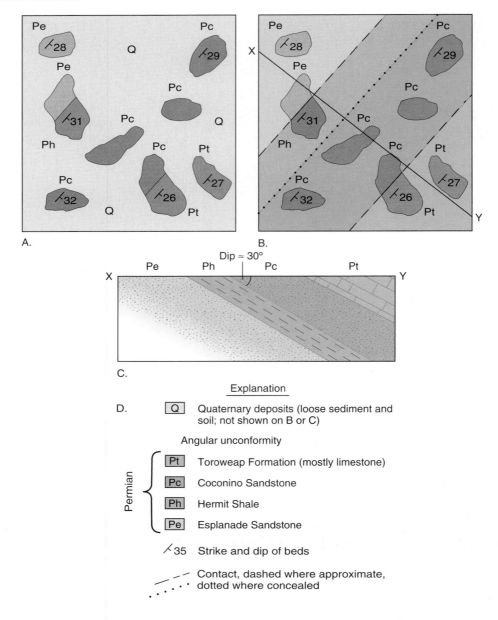

FIGURE 15.1

Making a geologic map. See text for discussion. A. Outcrop map. B. Contacts drawn in. C. Cross section along line *X-Y.* D. Explanation, with rock units and angular unconformity listed chronologically.

correlated, as illustrated in Figure 15.1B. Identical contacts are joined from one outcrop to another, using dashed lines to indicate they are not actually visible. In some places, no contacts are actually exposed, but their presence can be inferred. For example, say one outcrop is sandstone and an adjacent one is shale; the two must be in contact someplace, but the contact is buried. In such places, a "best-guess" dotted line is drawn to represent the contact (Fig. 15.1B).

One of the best ways to visualize what is shown on the map, and to double-check an interpretation, is to draw one or more geologic cross sections. The cross section in Figure 15.1C was drawn along the line *X-Y* on the map. That particular cross section was chosen because it best illustrates the overall structure. Note that the dip and strike symbols tell the angle and direction in which the layers are inclined at the surface. Correlation of the units below the surface completes the structure. On many

maps, the Quaternary units form a thin layer of soil and loose sediment resting on the underlying bedrock. In Figure 15.2, they are sediments deposited in a river valley by a river. Because they are often thin and unrelated to the main geologic structures, they are often not drawn on cross sections (Fig. 15.1).

Finally, rock units and unconformities are listed, with the oldest at the bottom and the youngest at the top, in the *Explanation* (Fig. 15.1D).

Appearance of Geologic Features from the Air and on Maps

You learned what common geologic structures look like in simple block diagrams in Chapter 14. Those diagrams illustrate ideal structures without the complexity of topography on the top surface. Even so, outcrop patterns of rock units seen on maps or from the air do resemble what you saw on the upper surface of the block diagrams. Hills and valleys complicate the smooth and regular appearance of formation contacts somewhat, but the basic pattern is there. The purpose of this section is to help you interpret what you see from an airplane or on a geologic map.

Horizontal Strata

If horizontal strata are exposed on a horizontal surface, only one formation will be seen. However, if the surface is not flat and erosion has cut into two or more horizontal formations, the contact lines between adjacent formations will parallel the contour lines. (Recall that contour lines show where horizontal planes, such as potentially defined by a rising sea level, intersect the

FIGURE 15.2

Horizontal Permian strata in the Flint Hills of northeastern Kansas. The geologic map and block diagram in A represent the area covered by the aerial photographs (scale 1:20,000) in B. *Q* on the map represents Quaternary deposits, and *Pmw, Psf, Pbc, Peb, Psb,* and *Peg* are symbols for Permian formations. The white bands in the photographs (B) are limestone; the gray bands are shale.

A.

B.

FIGURE 15.3

Inclined Cretaceous strata along Owl Creek, Hot Springs County, Wyoming. Ridges are sandstone, and valleys are shale. The geologic map and block diagram in A represent the area covered by the aerial photographs (scale 1:48,500) in B.

topography [Chapter 6]). Like contours, contacts bend so as to point upstream when crossing valleys, as shown in Figure 15.2. In an area with little relief, few strata will be exposed at the surface. The greater the relief, the deeper the land surface cuts into the underlying strata, exposing more of them. Because of differential erosion, resistant beds have steep slopes (narrow expression in map view), and easily eroded beds are marked by gentle slopes (wide expression in map view). If beds are not quite horizontal—that is, the dip is very low—contacts cross contour lines at small angles. In other words, *the flatter the dip, the more closely geologic contacts parallel contour lines.* Similarly, *the steeper the dip, the greater the divergence between contacts and contours.*

Inclined Strata

Strata inclined in one direction appear as roughly parallel bands on a map or aerial photograph, as shown in Figure 15.3. If the relief is moderate, the bands will be essentially parallel to the strike of the beds. In general, with increasing steepness of dip, the outcrop bands will be straighter and the contacts more regular, and the contacts will more closely parallel strike. Differential erosion may produce ridges and valleys that parallel the units. The direction of dip can be determined by the Rule of Vs, as described later, or in some places by differences in hill slope on opposite sides of ridges. Hills sloping in the direction of dip generally are gentler than those cutting across bedding (Fig. 15.3B).

Once the direction of dip is known, relative ages can be determined, because *older beds dip under (or toward) younger beds* (assuming beds are not overturned [Fig. 15.3A]).

Folded Strata

Complete folds are rarely exposed at the surface. Instead, erosion removes the crests of anticlines, leaving behind outcrop patterns that resemble the top surfaces of the block diagrams of Chapter 14 (Fig. 14.9, 14.10). If the folds are plunging, like those shown in Figure 15.4, the eroded, folded strata appear as winding or curved bands on geologic maps or aerial photographs. If non-plunging, the folded strata present roughly parallel bands, like those in Figure 15.3. Differential erosion commonly

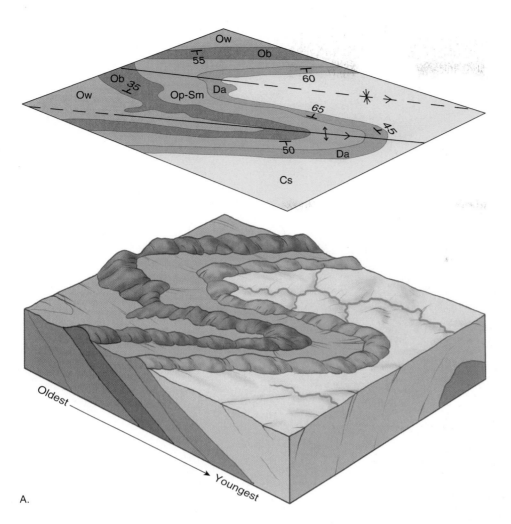

produces ridges and valleys that mimic the outcrop patterns.

As you learned from Problem 7 in Chapter 14, several things are worth remembering when observing eroded folds *in map view*.

1. When anticlines or domes are eroded, older rocks appear in the center and are surrounded by younger rocks.

2. When synclines or basins are eroded, younger rocks appear in the center and are surrounded by older rocks.

3. The contacts between beds in a plunging anticline, as seen in map view, bend so as to "point" in the direction of plunge.

4. The contacts between beds in a plunging syncline, as seen in map view, bend so as to "point" in the opposite direction from which it plunges.

Faulted Strata

Faults are shown on geologic maps with thick, black lines and symbols indicating the nature of the fault, if that is known (see symbols key, Fig. 14.4, in the previous chapter). Faults generally offset beds and contacts and thus commonly juxtapose quite different kinds of rock. Differential

FIGURE 15.4

Folded strata in the Ouachita Mountains, Atoka County, Oklahoma. Ridges are Ordovician (*Ob*) and Devonian (*Da*) chert; low areas are Ordovician, Silurian, and Carboniferous shale (*Ow, Op, Sm, Cs*). The geologic map and block diagram in A represent the area covered by the aerial photographs (scale 1:17,000) in B. Note that without strike and dip or relative age information it is difficult to tell which fold is a plunging anticline and which is a plunging syncline.

erosion may produce a cliff on one side of the fault or a narrow valley marking the fault itself. Fault valleys typically appear as straight, dark lines on aerial photographs, because the vegetation along them is commonly more lush than that surrounding them. On a geologic map, contacts, formations, and structures may end or be offset at faults.

Steeply dipping faults have straight traces (*trace* is the intersection between the fault plane and the surface) on maps or aerial photographs, whereas gently dipping faults may have irregular traces, depending on the topography. Faults dip under their hanging walls; therefore, on a geologic map, the direction of dip is toward the hanging wall. In map view, *normal faults* (Fig. 15.5) have older rocks on the footwall side. In contrast, *reverse faults* have younger rocks on the footwall side. This is because *erosion of the topographically higher side (the upthrown side) eventually exposes deeper and (usually) older rocks* (see Fig. 14.11). The traces of *strike-slip faults* are straight because of their common near-vertical dips. They are more likely than the other types of faults to be marked by straight, narrow valleys.

Thrust faults may have very irregular traces because of their low dips, often only

A.

B.

FIGURE 15.5

Basalt flows of Tertiary age (*Tb*) cut by normal faults, Sandoval County, New Mexico. Displacement ranges from near zero to about 50 m. Fault blocks are tilted eastward. The geologic map and block diagram in A represent the area covered by the aerial photographs (scale 1:31,680) in B. *Qal* indicates Quaternary alluvium, *Pz* (in block diagram) indicates rocks of Paleozoic age.

a few degrees. They are recognizable on a geologic map when structures above and below the fault do not correspond (for example, formations are not the same, or strikes and dips differ), and when the rocks in the hanging wall are older than those in the footwall. See Figure 14.11.

Unconformities (see also Chapter 13)

Angular unconformities (Fig. 13.6A) are recognized most easily on aerial photographs and geologic maps by nonparallel layers in rocks above and below the unconformity, as shown in Figure 15.6. Angular unconformities with low dips appear similar to low-angle thrust faults, except that rocks above the unconformity are younger than those below it.

Nonconformities (Fig. 13.6D) separate nonbedded, crystalline rocks from younger sedimentary rocks. They can be confused with intrusive igneous contacts. In nonconformities, the igneous contact is usually with the same formation throughout the map area, whereas intrusive contacts may cut across more than one formation. Relative ages established in the field are most important; igneous rocks are younger than the rocks they intrude, but older than rocks

A.

B.

FIGURE 15.6

Angular unconformity between nearly flat-lying Quaternary basalt lava flows (*Qcb*) and gently dipping Jurassic (*J*) and Cretaceous (*Km1, Km2,* and *Km3*) sedimentary rocks, Santa Fe and Sandoval Counties, New Mexico. The red lines on the map are dikes; *Qtg* is Quaternary gravel. The geologic map and block diagram in A represent the area covered by the aerial photographs (scale 1:54,000) in B.

EXPLANATION

Alluvium

Light-red wind-deposited sand and silt on benches and mesa tops, reworked in part by water; Recent valley fill and stream deposits.

UNCONFORMITY

Fanglomerate

Poorly sorted, in places rudely bedded, sand and angular fragments and boulders derived from older formations; somewhat indurated.

UNCONFORMITY

Burro Canyon formation

White, gray, and red sandstone and conglomerate with interbedded green and purplish shale.

Morrison formation

Variegated shale and mudstone; white, gray, rusty-red, and buff sandstone; rusty-red conglomerate; local thin limestone beds. At the top the Brushy Basin shale member, Jmb, consisting largely of bentonitic shale but including some sandstone and conglomerate lenses, and at the base the Salt Wash sandstone member, Jms, with more numerous and thicker sandstone beds.

Summerville formation

Thin-bedded red, gray, green, and brown sandy shale and mudstone.

Entrada sandstone and Carmel formation undivided

Orange, buff, and white, fine-grained, massive and cross-bedded Entrada sandstone at the top. Red sandstone and mudstone of the Carmel formation at the base.

UNCONFORMITY

Navajo sandstone

Buff and gray crossbedded, fine-grained sandstone.

Kayenta formation

Irregularly bedded, red, buff, gray, and lavender shale, siltstone, and fine- to coarse-grained sandstone.

Wingate sandstone

Fine-grained reddish-brown, cliff-forming sandstone, thickbedded, massive and crossbedded.

Pleistocene and Recent — QUATERNARY

Lower Cretaceous — CRETACEOUS

Upper Jurassic

Middle Jurassic — San Rafael group — JURASSIC

Glen Canyon group — JURASSIC (?)

Chinle formation

Red to orange-red siltstone, with interbedded lenses of red sandstone, shale, and limestone-pebble and clay-pellet conglomerate. Lenses of quartz-pebble conglomerate and grit at base.

UNCONFORMITY

Moenkopi formation

Chocolate-brown ripple-bedded shale, brick-red sandy mudstone, reddish-brown and chocolate-brown sandstone, and purple and reddish-brown arkosic conglomerate. Local gypsum beds. The upper member, Ɫmu, consisting of thin- and ripple-bedded shale with interbedded sandstone; the middle member, Ɫmm, consisting of ledge-forming beds of shale, sandstone, and arkosic conglomerate; and the lower member, Ɫml, consisting of poorly sorted sandy mudstone and local gypsum beds near base.

Cutler formation

Maroon, red, mottled light-red, and purple conglomerate, arkose, and arkosic sandstone. Thin beds of sandy mudstone.

UNCONFORMITY

Gneiss, schist, granite, and pegmatite

Upper Triassic

Lower Triassic — TRIASSIC

PERMIAN (?)

PRE-CAMBRIAN

Contact
Dashed where approximately located.

Indefinite contact
Includes inferred contacts and indefinite boundaries of surficial deposits.

D
U

Fault
Dashed where approximately located. U, upthrown side; D, downthrown side.

20

Strike and dip of beds

7500

Structure contours
Drawn on top of Entrada sandstone; dashes indicate projection above surface. Contour interval 100 feet. Datum is mean sea level.

Adit

x

Prospect

Mine

Source: U.S. Geological Survey.

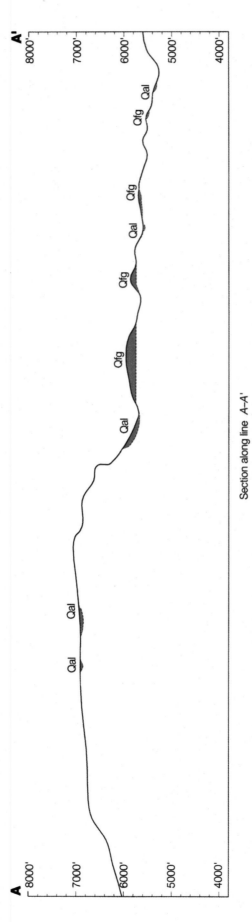

Section along line A–A'

FIGURE 15.13

Cross section along A–A' of Gateway Quadrangle, Colorado, geologic map, for use with Problem 3c. Qal and Qfg are already located for you.

c. Construct a geologic cross section from A to A' using the topographic profile in Figure 15.13 as a basis. The vertical scale and horizontal scale are the same, one inch to 2000 feet, so there is no vertical exaggeration. The Qal and Qfg are surficial sediment deposits; they are already on the profile (Fig. 15.13).

d. What is the approximate thickness of the Mesozoic rock units?

e. Notice that one mine and a number of prospects (holes or pits dug to investigate for ore, which are shown with an "X" on the map) are located in the area. The elements of interest are uranium, vanadium, and radium. If you were to explore this area for minerals containing these elements, on what formation(s) would you concentrate your efforts? (Evidence is on the map.)

4. Refer to Figure 15.14, a geologic map of the Black Hills of South Dakota, and to Figure 15.15, a composite *Landsat* image.

a. Study the outcrop pattern on the geologic map and the satellite image. What type of structure is present?

b. Based on the overall geologic structure and the map explanation, what is the age of the oldest rock in the structure?

c. Why are the areas underlain by the Minnelusa Sandstone and the Minnekahta Limestone—Opeche Formation wider on the west side of the structure than on the east?

d. Are the Cenozoic igneous intrusions older or younger than the surrounding rocks? How could you tell this if you weren't given the age of the rocks?

e. List the *sedimentary* units from the surface downward that would be penetrated by a well drilled at Edgemont in the southwestern part of the map.

f. The big reddish area on the *Landsat* image more or less coincides with the heart of the Black Hills. Why is it reddish? (See Chapter 7.)

g. Note the similarity in shape or pattern of the reddish areas of the satellite photo and some of the contact lines on the geologic map. This is due in part to elevation differences—that is, the elevation is higher within the red area, where rainfall is more plentiful—but that is not the only explanation. Examine the satellite image closely, compare with the geologic map, and suggest another possibility.

h. What causes the narrow southeast-trending bright red streaks on the east side of the satellite image?

EXPLANATION

SEDIMENTARY ROCKS

White River group
Brule clay and Chadron formation
Gravel at base

UNCONFORMITY

IGNEOUS ROCKS

Porphyries, grorudite, phonolite
quartz monzonite
Rhyolite tuff and obsidian
Stocks, sills, and dikes

UNCONFORMITY

SEDIMENTARY ROCKS

Hell Creek formation
Sandstone and shale; coal
Fox Hills sandstone
Lower limit variable; local coal

Pierre shale

Niobrara formation
Impure chalk and limy shale

Carlile shale

Greenhorn limestone

Belle Fourche and Mowry shales
Dark shale including sandstones

Skull Creek shale and
Inyan Kara group
Fall River sandstone (Dakota Sandstone)
Fuson shale, Minnewaste limestone
and Lakota sandstone

Morrison formation
Shale, mostly gray, sandstone and limestone

UNCONFORMITY

Sundance formation
Greenish shale, buff and red sandstone,
overlain locally by Unkpapa sandstone

Spearfish formation
Red sandy shale and sandstone;
gypsum members
"The Red Valley"

Minnekahta limestone and
Opeche formation
Red sandy shale

Minnelusa sandstone
Gray, red, and buff sandstones, mostly limy;
red shale at base

Pahasapa limestone and
Englewood limestone

UNCONFORMITY

Whitewood limestone
Hard, massive, buff limestone
in northern Black Hills

UNCONFORMITY

Deadwood formation
Gray to red sandstone and greenish shale,
in part glauconitic, slabby limestone,
flat-pebble limestone conglomerate,
quartz conglomerate at base,
thins to south

UNCONFORMITY

METAMORPHIC ROCKS

Amphibolite
Metadiorite and metagabbro; many
small dikes not shown

Schists, slates, and quartzites

Sandstone, arkosic, grit,
and conglomerate
Partly quartzite

UNCONFORMITY

IGNEOUS ROCKS

Granite and pegmatite
Some granite is gneissoid; granite has
many inclusions of schist

— — — — FAULT

SCALE
1 0 1 2 3 4 5   10 Miles

After N. H. Darton, 1951
Geologic Map of South Dakota
U. S. Geol. Survey

CENOZOIC (TERTIARY)

MESOZOIC

PALEOZOIC

PRECAMBRIAM

FIGURE 15.14
Geologic map of the Black Hills, South Dakota, for Problem 4.
Source: South Dakota Geological Survey.

FIGURE 15.15
Composite false-color *Landsat* satellite image of the Black Hills, South Dakota/Wyoming, for Problem 4.

5. Figure 15.16 is a geologic map of an area that has undergone faulting, folding, igneous activity, and several periods of erosion. Using the Rule of Vs, place strike and dip symbols at several places on the map to help you visualize the structures.

   a. What is the general strike and dip (just directions, not numbers) of the gabbro dike in the northeast corner? How do you know?

   b. Is the gabbro older or younger than fault *A*? How do you know?

   c. Is the dip of fault *B* greater or less than fault *A?* How do you know?

   d. In what direction does fault *A* dip? How do you know?

   Which side of fault *A* is the hanging wall and which the footwall? How do you know?

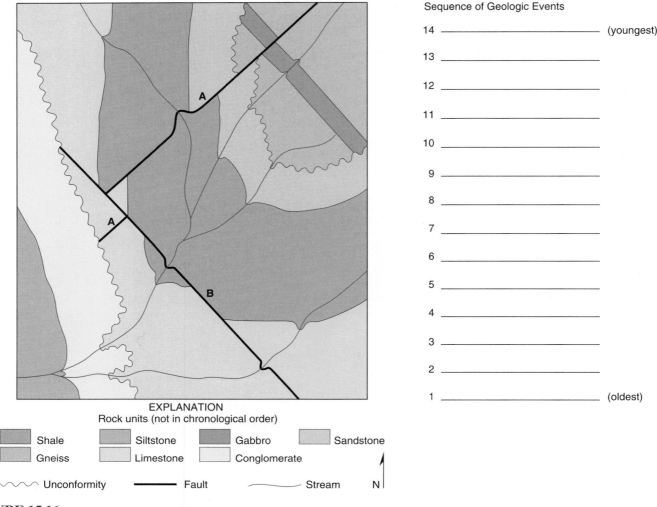

Sequence of Geologic Events

14 _____ (youngest)

13 _____

12 _____

11 _____

10 _____

9 _____

8 _____

7 _____

6 _____

5 _____

4 _____

3 _____

2 _____

1 _____ (oldest)

EXPLANATION
Rock units (not in chronological order)

Shale   Siltstone   Gabbro   Sandstone

Gneiss   Limestone   Conglomerate

〰〰 Unconformity   ——— Fault   ⌒ Stream   N↑

FIGURE 15.16
Geologic map for use with Problem 5.

e.   Is fault *B* older or younger than fault *A?* How do you know?

Which side of fault *B* is the hanging wall and which the footwall?

f.   Is the unconformity between the limestone and conglomerate an angular unconformity, a disconformity, or a nonconformity? How do you know? What type of unconformity is the one between the gneiss and the sandstone?

g.   In the spaces to the right of the map, list the sequence of geologic events that produced the geologic relations shown on the map. Begin with the oldest.

6.   Figure 15.17 is part of a geologic map of the Tazewell Quadrangle, Tennessee. The area is part of the Appalachian Mountain system, which here consists of Cambrian and Ordovician sedimentary rocks folded and cut by two prominent faults, the Wallen Valley thrust fault and the Hunter Valley thrust fault (extreme southeast corner).

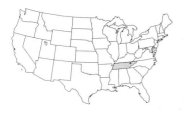

Scale 1:24,000
C.I. 20 feet

FIGURE 15.17

(Pages 300–301) Geologic map of portion of Tazewell Quadrangle, Tennessee, for use with Problem 6. Not all formations listed in the Explanation appear on this portion of the map. Source: U.S. Geological Survey.

a. What is the general structure of the rocks between the Wallen Valley (center of map) and Hunter Valley (southeast corner) thrust faults? That is, are the strata horizontal, inclined, folded?

b. Why are the contacts between €cr, Oc, Ol, and On wavy, whereas those farther southeast, such as between Ow and Ohb, are smooth?

c. In what direction do the two faults dip? Based on the outcrop pattern of the faults, is the dip steep or gentle? Compare the ages of the rocks immediately adjacent to each of the two faults. Do older rocks overlie or underlie younger rocks along the faults?

d. The outcrop pattern northwest of the Wallen Valley thrust fault is different from that to the southeast. Why? What features are present? Sketch an approximate cross section from the W in the word Wallen northwest to the edge of the map. You need not show the topography nor be too accurate in locating contact lines; the purpose is to see whether you understand the general structure.

e. Depressions, as indicated by depression contours, are abundant in some formations and absent in others. Explain. (Hint: In what kind of rock do the depressions occur?)

f. Draw a line on the map along which a cross section that best shows the geologic structures could be drawn. Label it A on one end and B on the other.

GEOLOGIC QUADRANGLE MAP
TAZEWELL QUADRANGLE, TENNESSEE
GQ-465

Contact
*Dashed where approximately located;
dotted where concealed*

Fault
*Dashed where approximately located; dotted where
concealed. U, upthrown side; D, downthrown side;
T, upper plate of minor thrust fault*

Tear fault
*Showing relative movement. Dashed where approxi-
mately located; dotted where concealed*

Major thrust fault
*Dashed where approximately located; dotted where
concealed*

Anticline
*Showing trace of axial plane; dashed where approxi-
mately located; dotted where concealed*

Syncline
*Showing trace of axial plane; dashed where approxi-
mately located; dotted where concealed*

Strike and dip of beds

Strike and dip of overturned beds

Horizontal beds

Quarry

Fe
Occurrence of limonite

Zn
Zinc prospect pit

CAMBRIAN

On — Newala Dolomite

Ol — Longview Dolomite

Chepultepec Dolomite
Ocu, upper member
Ocl, lower member
(Ocu / Oc / Ocl)

Copper Ridge Dolomite — €cr

Maynardville Formation
€mnc, Chances Branch Dolomite Member
€mnl, Low Hollow Limestone Member
(€mnc / €mnl / €mn)

Nolichucky Shale — €n

Maryville Limestone — €m

Rogersville Shale — €rv

Rutledge Limestone — €rt

Pumpkin Valley Shale — €pv

Rome Formation
€rd, dolomite member
(€rd / €r)

Knox Group

Consauga Group

Lower Ordovician

Upper Cambrian

Middle Cambrian

Lower Cambrian

QUATER-NARY

SILURIAN

ORDOVICIAN

Qal — Alluvium

Sct — Clinton Shale
Not exposed but inferred in subsurface

Clinch Sandstone
Scp, Poor Valley Ridge Member
Sch, Hagan Shale Member
(Scp / Sc / Sch)

Os — Sequatchie Formation

Or — Reedsville Shale

Ot — Trenton Limestone

Oe — Eggleston Limestone

Ohc — Hardy Creek Limestone

Obh — Ben Hur Limestone

Ow — Woodway Limestone

Ohb — Hurricane Bridge Limestone

Omc — Martin Creek Limestone

Orc — Rob Camp Limestone

Op — Poteet Limestone

Od — Dot Formation

Chickamauga Group of Swingle (1964)

Middle Silurian

Lower Silurian

Upper Ordovician

Middle Ordovician

7. a. Go to *http://geology.about.com/od/stategeologicmaps/* (or link to it through *www.mhhe.com/jones6e*—see Preface), and use the geologic map of Oklahoma and the cross sections to answer the following questions.

    (1) What is the age (give the name of the period) of the pre-Quaternary bedrock exposed near Oklahoma City? Near Tulsa?

    (2) Why do the Quaternary sediments (in yellow) form streaks across the state?

    (3) What is the age of the rocks exposed in the Wichita Mountains and what kinds of rocks are they?

    (4) The Wichita Mountains and the Anadarko Basin are separated by a steep fault. Which side has gone down relatively?

        Is the fault a normal or reverse fault according to the cross section?

    (5) What type of contact (see Chapter 14) separates the Precambrian and Cambrian igneous rocks from the Cambrian sedimentary rocks in the Anadarko Basin?

    (6) Many of the faults in the Ouachita Mountains are thrust faults that originally had low dips. The dips in cross section B-B' appear steeper than they really are, because the cross sections are vertically exaggerated about 10 ×. Based on cross section B-B', in which relative direction were the rocks above the thrust-fault surfaces moved, toward the north or south?

        Rocks of the Ouachitas have also been folded. Based on B-B', did folding occur before or after thrust faulting?

    b. You can find information about the geology of your state on the web at *http://geology.about.com/od/ stategeologicmaps/geomapsusstates/*, or go to *www.consrv.ca.gov/cgs/index.htm* and link to your state geological survey to see what's available, or try some of the universities or colleges in your state.

In Greater Depth

8. Figure 15.18 is a geologic map of the Wetterhorn Peak Quadrangle, Colorado. This area is within the San Juan Mountains of Colorado, a range composed principally of Tertiary volcanic rocks. Tertiary units present are: (1) San Juan Formation (*Tsj* in gold), which consists of massive tuff, volcanic breccia, and lava flows of rhyolitic to andesitic composition; (2) porphyritic quartz latite, a fine-grained rhyolite-like rock with both plagioclase and K-feldspar (*Tql* in red); and (3) Potosi Volcanic Group, consisting mainly of welded pyroclastic-flow tuff of rhyolitic composition (*Tp* in green).

    a. Is the porphyritic quartz latite (the red bodies) an extrusive or intrusive igneous rock? Explain. What geologic term is given to the long, narrow bodies of *Tql?*

    b. List the three Tertiary units in order of age, with youngest first.

Scale 1:24,000
C.I. 40 feet

0                    $\frac{1}{2}$ 1 Mile

0   .25   .5   1 Kilometer

FIGURE 15.18
(Page 304) Geologic map of portion of
Wetterhorn Peak Quadrangle, Colorado, for
use with Problem 8.
Source: U.S. Geological Survey.

c. Why do the long, narrow bodies of *Tql* follow such straight lines, even though the terrain is mountainous?

d. Notice that the narrow bodies of *Tql* radiate outward from the irregular-shaped bodies of *Tql*, but the pattern is not random; certain orientations or strikes are preferred. Suggest one or more reasonable hypotheses to explain the preferred orientations.

e. Five types of Quaternary surficial deposits are present on the map. All have *Q* as the first letter of their symbol. Two originated as stream deposits, the others as accumulations at the base of a cliff or steep slope, glacial deposits, or landslide deposits. Locate the five types on the map, and based on their locations, speculate on the origin for each, using its symbol to identify it.

f. The red-dot pattern on Bighorn Ridge symbolizes a type of alteration, commonly associated with orebodies, that is caused by circulation of hydrothermal (hot water) solutions. Why might the alteration have occurred at that particular locality?

9. Refer to Figure 15.12 and answer the following questions.

a. What kind of contact separates Precambrian rocks from the Cutler Formation? Be specific. Based on the nature of its lower contact, its stratigraphic position, and the description in the *Explanation,* give a hypothesis for the origin of the Cutler Formation.

b. Note that the Navajo Sandstone is present in the southwest part of the map but does not continue to the north. Why? What kind of contact separates the Kayenta and Carmel Formations where the Navajo is absent? Be specific.

c. The red lines running diagonally across the map from northwest to southeast are called *structure contours* because they are drawn on the top surface of a particular formation and show the shape of that buried surface. On this map, they are drawn on the top of the Entrada Sandstone (notice that they are dashed where the present-day surface is below the top of the Entrada). What is the strike of that surface? What is the gradient of that surface in feet per mile? Use trigonometry to calculate the angle of dip of the surface. (Hint: Convert the numerator and denominator of the gradient to the same units, then divide. The result is the *tangent* of the dip angle. The dip angle is the *arctangent* or *inverse tangent* on most calculators. If necessary, see your lab instructor for help.) Measure the dip on your cross section and compare it with the calculated dip.

CHAPTER 16

Earthquakes

Materials Needed

- Pencil and eraser
- Ruler
- Drawing compass, or pencil on string, to draw circles
- Calculator

Introduction

Earthquakes are both destructive and informative. Energy from them travels through and around the Earth as three types of waves, called *P, S,* and surface waves. In this lab, you will learn to identify the three types of earthquake waves on a seismogram and to use the seismogram to locate an earthquake and determine its magnitude. You will also learn how the intensity or degree of shaking of an earthquake is measured and mapped and how intensity depends on the characteristics of the ground being shaken. Finally, a simple example will show how earthquakes are used to examine the Earth's interior.

Earthquakes are vibrations resulting from sudden movements within the Earth. They are most common at tectonic plate boundaries, as discussed more fully in Chapter 17. Most earthquakes result from slippage along faults, but some are caused by other kinds of movement. For example, the upward movement of magma into Mount St. Helens in 1980 caused earthquakes that first alerted geoscientists to the probability of an eruption. Huge landslides and collapse of caves can produce earthquakes, and vibrations caused by explosions, such as underground nuclear tests, are very similar to natural earthquakes. We

can only presume that those who named the dinosaur *Seismosaurus* thought that, in its day, the huge dinosaur's footsteps must have generated some excitement.

Seismology is the study of earthquakes, and the branch of seismology known as earthquake seismology focuses specifically on the causes and effects of earthquakes, especially as earthquakes affect people. Because large earthquakes may result in death and destruction, a goal of earthquake seismologists is to be able to predict earthquakes. Other seismologists make use of the energy given off by earthquakes (or explosions) to see within the Earth, just as we process light energy with our eyes or sound energy with our ears to form images of our surroundings.

Seismic Waves

When a fault suddenly slips **(ruptures),** rocks on either side elastically snap back toward their original unstressed shapes. This elastic rebound releases a pulse of energy that radiates out in all directions (Fig. 16.1). The "point" below the surface at which rupture begins is called the **focus,** or **hypocenter,** of the earthquake. The focus is not just a point, but a small area along a fault that ruptures. Rupture may begin at one spot, but then propagates along the

fault in an irregular, jerky fashion. In fact, a series of smaller **aftershock** earthquakes commonly rattle the region for a while as stresses are relieved along the main fault or other nearby faults. The **epicenter** is the point on the Earth's surface directly above the focus. It is the epicenter that is first reported by the press, because it is of most concern to people and is easily determined.

Seismic energy is released in the form of **seismic waves,** which cause the Earth to vibrate or deform elastically as they pass by. The three principal kinds of seismic waves are as follows:

> *P* **wave (primary wave** or **compressive wave;** Fig. 16.2). The energy of a *P* wave travels through rock as a sequence of back-and-forth vibrations *parallel* to the direction of movement. Its passage causes alternate pushing (compression) and pulling (dilation) on particles in its path. *P* waves can travel in solids or liquids, and are the fastest of the three types of seismic waves.

> *S* **wave (secondary wave** or **shear wave;** Fig. 16.2). The energy of an *S* wave travels through rock as a sequence of vibrations *perpendicular* to the direction of movement. Its passage causes particles to vibrate up and

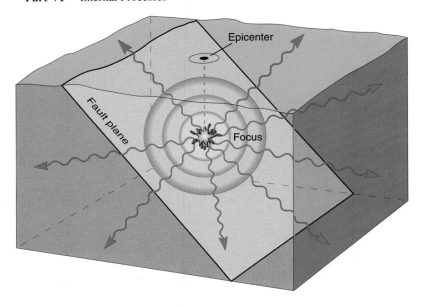

FIGURE 16.1
An earthquake originates at its *focus,* below the surface. The *epicenter* lies directly above the focus, on the surface. Seismic waves radiate outward in all directions from the focus.

A. P wave

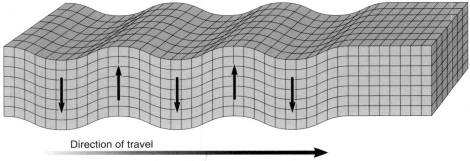

B. S wave

FIGURE 16.2
A. *P* waves vibrate back and forth, parallel to the direction of travel. B. *S* waves vibrate perpendicular to the direction of travel.

down or sideways. *S* waves can travel in solids but not in liquids, and their velocity is between that of the *P* and surface waves.

Surface waves. Surface waves travel along the surface of the Earth as two types of waves. One of these (called a Love wave) has a horizontal shearing motion similar to the motion of a snake across the ground, while the other (a Rayleigh wave) has a rolling up-and-down motion much like water waves. Surface waves are the slowest seismic waves, but typically these cause the most damage.

Seismograms

Earthquakes are very common; hundreds of thousands are detected each year. However, most are too small or too remote to be felt, and relatively few cause damage or loss of life. Earthquakes, even very small ones, are recorded with **seismographs.** A worldwide network of seismographs constantly monitors earthquake activity and provides information about an earthquake within minutes of its occurrence.

The record obtained with a seismograph is a **seismogram.** One type of seismogram is recorded on a rotating drum as a series of wiggly lines (Fig. 16.3). Tick marks or breaks in the lines are precisely spaced and represent time elapsed; actual time is Universal Coordinate (or Greenwich Mean) Time and, for most stations, is accurate to a few thousandths of a second. Figure 16.3 shows a simple seismogram recording the *P, S,* and surface waves.

A seismograph station may have three seismographs arranged to record the vertical component and two horizontal components of motion (for example, E-W and N-S). The three together allow one station to give an analysis of an earthquake.

Locating Earthquakes

The distance from an earthquake focus can be calculated if the seismic-wave velocity and the time it took to travel to the seismograph are known. Earthquake data from seismograph stations are combined with data on how fast seismic waves travel through specific rock units to produce

FIGURE 16.3
Simple seismogram with *P, S,* and surface waves. Blue tick marks indicate time. Originally written on a rotating drum, time moves forward from left to right in minutes, then skips down to the next line. One seismic event stands out above the background noise. Modern seismic data are recorded digitally.

travel-time curves. **Travel-time curves** show how the time required for various types of waves to travel a given distance varies with distance. In Figure 16.4, curves are shown for a *P* and an *S* wave. If the times at which an earthquake occurred and the *P* wave arrived are known, the distance from the earthquake can be read directly from the graph.

But what about the more common case in which the time at which the earthquake occurred is unknown? Because the farther a *P* wave travels, the more the *S* wave falls behind, the *difference* in *P*- and *S*-wave arrival times can be used to obtain the distance to an earthquake[1]. This is seen as an increasing vertical space between travel-time curves in Figure 16.4. The distance to the earthquake epicenter can be determined from a seismogram by measuring the *difference* in *P*- and *S*-wave arrival times and finding the place on the travel time curves where they have the same time gap. The corresponding position on the horizontal axis gives the distance to the earthquake.

For example, let's say the seismogram in Figure 16.3 was recorded in Pierre, South Dakota, and the difference in arrival times of *P* and *S* waves is about 70 seconds, which on Figure 16.4 corresponds to a distance from the focus of about 500 km.

Although the distance to the earthquake focus is known, the direction from the seismograph station to the focus is not known. Thus, on Figure 16.5, a circle of radius 500 km is drawn around Pierre. The earthquake, if a shallow one, must be located someplace on that circle. But where? One circle is not enough. Three intersecting circles for the same earthquake from three seismograph stations are needed, as

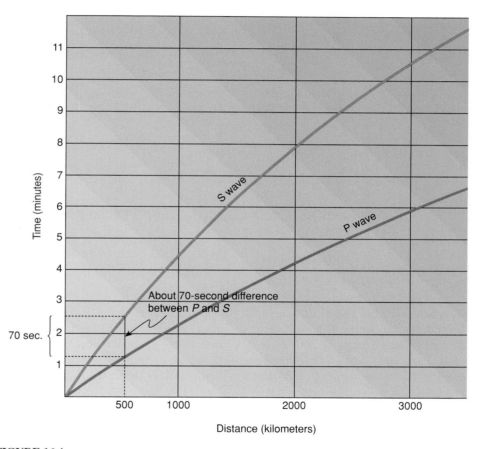

FIGURE 16.4
Travel-time curves for *P* and *S* waves. Note that the time interval between curves increases with distance. The seventy-second difference in arrival time between *P* and *S* waves indicates a distance to focus of about 500 km.

[1]*The process is analogous to one you may have used to figure out how far away lightning was during a thunderstorm. You see a lightning flash instantaneously because the velocity of the light waves is very fast (3×10^5 km/sec), but the thunder takes awhile to get to you because the velocity of sound is much slower (about 0.34 km/sec). The far-ther away the lightning, the longer it takes for the thunder to get to you. If you count the seconds between lightning and thunder and multiply by the velocity of sound—0.34 km/sec—you know how far away the lightning was. A difference of 4 seconds means a distance of 4×0.34 km = 1.36 km, about 0.85 mile.*

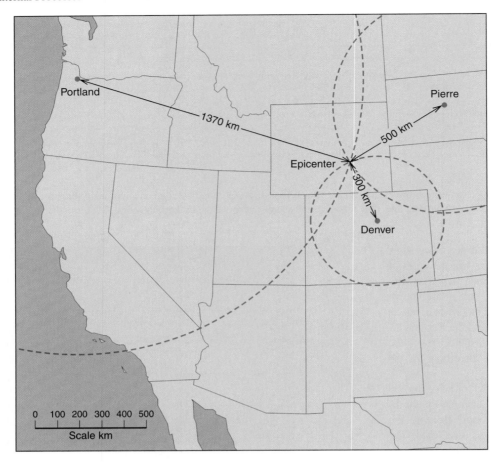

FIGURE 16.5
Seismograms from seismograph station at Portland, Oregon; Denver, Colorado; and Pierre, South Dakota are used to locate the epicenter near Douglas, Wyoming, with three intersecting circles.

illustrated in Figure 16.5. Thus a seismograph station in Denver, Colorado, reports the earthquake as occurring 300 km away while a station in Portland, Oregon, reports it to be 1370 km away. Circles drawn about all three stations intersect at the epicenter, which lies near Douglas, Wyoming (Fig. 16.5).

When did the earthquake occur? On Figure 16.4, dotted horizontal lines are extended to the time axis from the point of intersection between the vertical line at a 500-km distance and the P- and S-wave curves. From this it can be determined that it took the P wave about 1.3 minutes, and the S wave about 2.5 minutes, to get from the earthquake focus to Pierre, South Dakota. Although not shown in Figure 16.3, the actual time represented by each blue tick mark is known, so the actual arrival time of P and S waves can be determined. For example, let's say that the P wave arrived at 13:45.3 Universal Coordinate

Time and the S wave 1.2 minutes (70 seconds) later at 13:46.5. To find the time at which the earthquake occurred, subtract 1.3 minutes (the travel time for the P wave) from 13:45.3 to get 13:44 for the origin time of the earthquake. As a check, subtract 2.5 minutes (the travel time for the S wave) from 13:46.5 to get 13:44 again. Thus, the earthquake occurred at 13:44 Universal Coordinate Time. Since Douglas, Wyoming, is in the Mountain time zone, subtract 7 hours to get a local time of 6:44 A.M.

Earthquake Intensity and Magnitude

The size of an earthquake is described in two ways: intensity and magnitude. **Earthquake intensity** is a measure of the destructive effects of an earthquake at a particular place and is a qualitative mea-

surement based on people's experiences in an earthquake. Intensity is assessed with the *Modified Mercalli Scale* (Table 16.1), which has twelve categories (I to XII) that assess earthquake damage and human reactions to the quake. Questionnaires based on the scale are mailed out after an earthquake, and residents' responses are used to prepare a map of earthquake intensity. Lines of equal intensity, **isoseismals,** are like contours and outline the area where various levels of damage were recorded. Figure 16.6 illustrates the results for the Douglas, Wyoming, earthquake. The intensity felt throughout the area depends not only on the size and duration of the earthquake but also on the strength of the underlying ground. Areas underlain by loose sediment shook much more than those on solid bedrock. Notice that some people in Pierre felt the earthquake, but it was not felt in the surrounding area. These results illustrate why detailed earthquake intensity data

Table 16.1
Modified Mercalli Scale

I Not felt by people, except rarely under especially favorable circumstances.

II Felt indoors only by persons at rest, especially on upper floors. Some hanging objects may swing.

III Felt indoors by several people. Hanging objects may swing slightly. Vibration like passing of light trucks. Duration estimated. May not be recognized as an earthquake.

IV Felt indoors by many, outdoors by few. Hanging objects swing. Vibration like passing of heavy trucks; or sensation of a jolt like a heavy ball striking the walls. Standing automobiles rock. Windows, dishes, doors rattle. Wooden walls and frame may creak.

V Felt indoors and outdoors by nearly everyone; direction estimated. Sleepers wakened. Liquids disturbed, some spilled. Small, unstable objects displaced, some dishes and glassware broken. Doors swing; shutters, pictures move. Pendulum clocks stop, start, change rate. Swaying of tall trees and poles sometimes noticed.

VI Felt by all. Damage slight. Many frightened and run outdoors. Persons walk unsteadily. Windows, dishes, glassware broken. Knickknacks and books fall off shelves; pictures off wall. Furniture moved or overturned. Weak plaster and masonry cracked.

VII Difficult to stand. Damage negligible in buildings of good design and construction; slight to moderate in well-built ordinary buildings; considerable in badly designed or poorly built buildings. Noticed by drivers of automobiles. Hanging objects quiver. Furniture broken. Weak chimneys broken. Damage to masonry; fall of plaster, loose bricks, stones, tiles, and unbraced parapets. Small slides and caving in along sand or gravel banks. Large bells ring.

VIII People frightened. Damage slight in specially designed structures; considerable in ordinary substantial buildings, partial collapse; great in poorly built structures. Steering of automobiles affected. Damage or partial collapse to some masonry and stucco. Failure of some chimneys, factory stacks, monuments, towers, elevated tanks. Frame houses moved on foundations if not bolted down; loose panel walls thrown out. Decayed pilings broken off. Branches broken from trees. Changes in flow or temperature of springs and wells. Cracks in wet ground and on steep slopes.

IX General panic. Damage considerable in specially designed structures; great in substantial buildings, with some collapse. General damage to foundations; frame structures, if not bolted, shifted off foundations and thrown out of plumb. Serious damage to reservoirs. Underground pipes broken. Conspicuous cracks in ground; liquefaction.

X Most masonry and frame structures destroyed with their foundations. Some well-built wooden structures and bridges destroyed. Serious damage to dams, dikes, embankments. Landslides on riverbanks and steep slopes considerable. Water splashed onto banks of canals, rivers, lakes. Sand and mud shifted horizontally on beaches and flat land. Rails bent slightly.

XI Few, if any masonry structures remain standing. Bridges destroyed. Broad fissures in ground; earth slumps and landslides widespread. Underground pipelines completely out of service. Rails bent greatly.

XII Damage nearly total. Waves seen on ground surfaces. Large rock masses displaced. Lines of sight and level distorted. Objects thrown upward into the air.

---

are invaluable in preparing the earthquake hazard maps that guide property development and regional engineering and construction standards.

Earthquake magnitude is a quantitative measure of the amount of energy released by the earthquake. The original magnitude scale was the *Richter Magnitude Scale*. It defined magnitude as the logarithm (base 10) of the amplitude[2], measured in micrometers (thousandths of a millimeter), of the largest waves recorded on a specified seismograph at a distance of 100 km from the epicenter. The distance must be specified because amplitude decreases with distance from the focus; the seismograph is specified to account for instrument variability. The assumption was that more energetic earthquakes would produce bigger wiggles, so that the height—or

amplitude (top of Fig. 16.7)—of a selected wave would be proportional to the amount of energy released. There are no lower or upper limits to the Richter scale. However, the largest Richter-scale magnitude recorded thus far is 8.6. Because the Richter scale is a logarithmic scale, an increase of one unit in the scale corresponds to a 10-fold increase in amplitude. An increase in one unit in magnitude suggests an increase in the amount of energy released of about 32 times.

The original Richter scale was designed in 1935 for use with shallow-focus earthquakes occurring in southern California, where the largest waves are common *S* waves. Since then, similar amplitude-based earthquake scales have been developed for more distant and/or deeper earthquakes. However, it has been found that these

amplitude-based scales do not work well for the largest earthquakes, those with magnitudes greater than about 8.

Thus another kind of earthquake magnitude scale, called the **moment-magnitude scale,** is now preferred by many seismologists because it is directly related to the amount of energy released by fault rupture and it is accurate for earthquakes of any size. It is based on the strength of the rock in which the fault occurs, the area ruptured by the fault, and the average displacement on the fault. The moment magnitude can be determined by detailed analysis of the seismogram, which is the usual practice.

[2]*The logarithm (base 10) of a number is the power to which 10 must be raised to equal that number. A logarithm can be a whole or fractional number. For example, $log_{10}100 = 2$; $10^2 = 100$. And $log_{10}5 = 0.699$; $10^{0.699} = 5$.*

FIGURE 16.6

Intensity map for earthquake near Douglas, Wyoming, with isoseismal lines based on the Modified Mercalli scale. The contours are drawn as relatively simple, generally concentric shapes that are intended to summarize the most important patterns in the data. Intensely convoluted contours or contours encircling single data points are of little value.

Because the data necessary to calculate the moment magnitude require some time to gather, the first magnitude reported to the press might be an amplitude-based magnitude, which is later updated with the energy-based moment magnitude. Fortunately, the magnitudes of the amplitude-based scales are comparable to the moment-magnitude scale (at least up to about 8). Although the moment-magnitude scale is preferred, we focus on determining Richter magnitudes because the calculations are more straightforward.

Figure 16.7 illustrates the measurement of the Richter magnitude of an earthquake modified from the original Richter method (after Bolt, 1988). (1) Measure the amplitude of the largest surface wave in millimeters; in the example, it is 20 mm. (2) Use the graph to determine the Richter magnitude by running a straight-edge from the appropriate *Distance to epicenter* (assumed to be 60 km in the example) to the appropriate *Amplitude;* read the magnitude where this line crosses the *Magnitude* scale (about 5.6 in the example). This method applies only to local earthquakes less than 500 km away.

Using Seismic Waves to See Beneath the Surface

Seismic body waves (*P* and *S*) passing through the Earth do not follow straight-line paths but are bent (refracted) and reflected when they encounter rocks of contrasting densities (Fig. 16.8). Their behavior is just like light in this regard. When light travels from air (density A) to water (density B), some light is reflected while the rest is bent as it passes into the water. Figure 16.8 shows a *P* wave being reflected and traveling back to the surface. If the velocity and travel time of the *P* wave are known, the distance traveled can be calculated, and the depth to the reflecting horizon can be determined. Reflection seismology, using energy from human-produced explosions or vibrations, is used to map geologic structures below the surface and is one of the most important methods of exploration used in the oil industry.

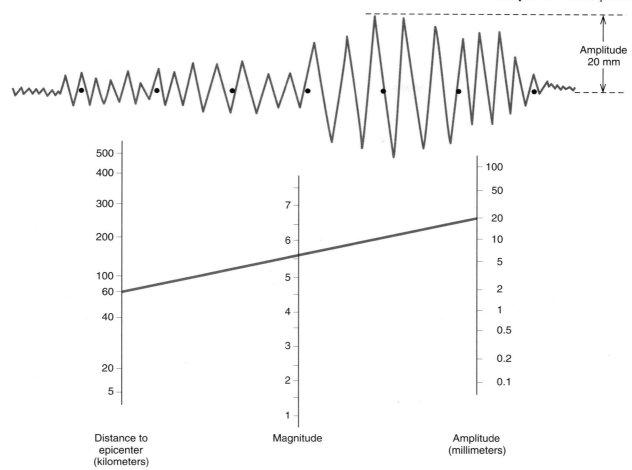

FIGURE 16.7
A line joining the amplitude of the largest wave on the seismogram and the distance to the epicenter crosses the magnitude scale at 5.6. This method is applicable to nearby earthquakes only.

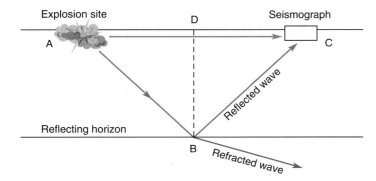

To determine depth to the reflecting horizon

1. Knowing the travel time and distance, ADC, from the explosion site to the seismograph, calculate the *P*-wave velocity (velocity = distance/time).

2. Knowing the travel time of the reflected wave and the velocity found in Step 1, calculate the distance, ABC, travelled by the reflected wave (distance = velocity × time).

3. Use the distances AD and AB to calculate the depth to the reflecting horizon, DB. (Remember the Pythagorean theorem? The square of the hypotenuse of a right triangle equals the sum of the squares of its legs—in this case, the hypotenuse is AB, and AD and DB are the legs, so $AB^2 = AD^2 + DB^2$.)

FIGURE 16.8
Reflection of a *P* wave and calculation of depth to reflector.

Hands-On Applications

The Loma Prieta Earthquake: Background Information

As millions of Americans were tuned in to the beginning of the third game of the World Series (championship baseball) at Candlestick Park, San Francisco, California, on October 17, 1989, a large earthquake occurred. It was the largest to strike the San Francisco Bay region, home to almost 6 million people, since the great earthquake of 1906, when fewer than a million people lived there. The earthquake was felt over 400,000 square miles and resulted in 67 deaths; 12,000 people were left homeless and it caused $6 billion in property damage. In terms of cost, it was one of the most expensive natural disasters ever in the United States, and it was the most costly earthquake since 1906.

The following problems illustrate how the epicenter was located, how the magnitude and intensities were determined, how the intensity was influenced by the nature of the surficial material, and how energy from the earthquake could be used to view the interior of the Earth. In each case, the scientific method is used to answer basic problems.

Objectives

If you are assigned all the problems, you should be able to:

1. Identify P, S, and surface waves on a simple seismogram.

2. Locate the epicenter of an earthquake using seismograms and travel-time curves.

3. Explain the basis for determining the magnitude and intensity of an earthquake, and draw isoseismals from intensity data.

4. Describe how different earth materials (mud, alluvium, bedrock) respond to seismic waves.

5. Use seismic reflection data to determine the depth to a reflector.

Problems

1. Figure 16.9 shows three simulated seismograms from the October 17, 1989, Loma Prieta Earthquake. Each was recorded at a different seismic station. Use them to answer the following:

   a. Where was the epicenter?

      (1) Find the differences in arrival time between P and S using Figure 16.9. Next determine the distance to the epicenter from each of the three stations using the travel-time curves in Figure 16.10.

      Distance from epicenter of: Station A

                                                     Station B

                                                     Station C

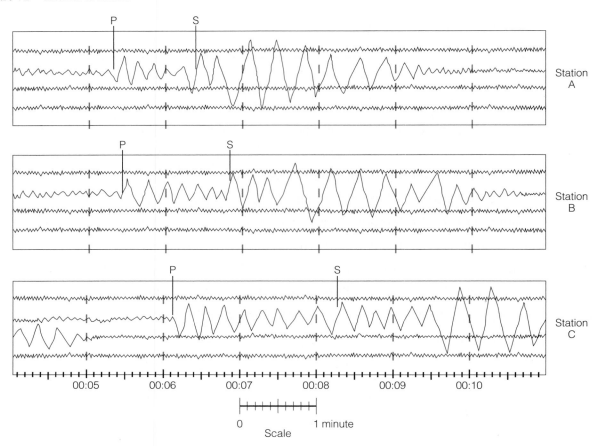

FIGURE 16.9

Three seismograms for Problem 1. Blue tick marks on the seismograms are one minute apart. The Universal Coordinate Times given below the lower seismogram also apply to the upper two seismograms (for example, the *P* wave arrived at Station A between 00:05 and 00:06). Because Universal Coordinate Time is based on a 24-hour day, 00:05 is the same as 12:05 a.m. What are the differences in arrival times for *P* and *S* waves at Station A? _____ Station B? _____ Station C? _____ Use these values in Figure 16.10 to obtain the distance to the epicenter.

FIGURE 16.10
Travel time curves for *P* and *S* waves. This enlargement of Figure 16.4 has the same time scale as the seismograms in Figure 16.9 and is to be used for Problem 1. What is the distance from the epicenter of Station A? _____ Station B? _____ Station C? _____.

    (2)   Plot the locations of the stations on the map in Figure 16.11. See the caption for Figure 16.11 for station locations.

    (3)   Locate the epicenter of the earthquake on Figure 16.11 from intersecting circles centered at the three stations. Note that your epicenter and your neighbor's may not be exactly the same. Why not?

  b. When did the earthquake occur?

    (1)   Use the travel-time curves and the time values indicated on the seismograms to determine the time of occurrence.

    (2)   The time you determined is Universal Coordinate Time. What time would it have been in the Pacific Daylight time zone if there is a seven-hour difference?

  c. How big was the earthquake? Determine the Richter magnitude from seismogram A using the method shown in Figure 16.7.

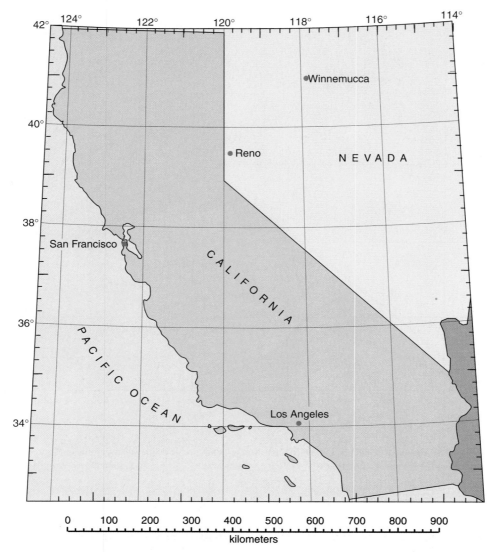

FIGURE 16.11

Map for Problem 1. Plot stations A (lat. 40°08′N, long. 121°30′W), B (lat. 34°00′N, long. 118°00′W), and C (lat. 41°00′N, long. 114°30′W). Then locate the epicenter using the distances from the epicenter obtained from Figure 16.10.

2. Questionnaires designed for assigning Modified Mercalli scale intensities to the October 17 earthquake were returned from the locations shown on Figure 16.12. The responses were divided into five groups, as shown in Table 16.2. Using these responses, assign Modified Mercalli Scale intensities to the five groups in Table 16.2 using Table 16.1 as a guide. Next, draw isoseismals on the maps based on intensity distribution. How does the area of highest intensity compare with the epicenter you located in Problem 1?

FIGURE 16.12
Map of San Francisco Bay area showing locations for which intensities have been assigned. Letters A through E correspond to earthquake intensities described in Table 16.2. For use with Problems 2 and 3.

Table 16.2
Earthquake Intensity Survey

Questionnaire Category Descriptions	Modified Mercalli Scale Equivalents
A. Felt by all, and people were extremely frightened. Drivers had great difficulty controlling vehicles. Some buildings collapsed, others very seriously damaged. Concrete bridge collapsed. Smell of gas in some areas. Ground cracked.	
B. Felt by all, most people were frightened. Drivers had some difficulty steering. Some reasonably well-built buildings partially collapsed, poorly built buildings badly damaged. Some water towers damaged. Large trees had tops broken. Some reports of furniture and people being thrown in air. Cracks in ground.	
C. Felt by all, some people knocked down. Noticed by some drivers. Damage to buildings moderate, except for those of poor construction. Windows broken, things knocked off shelves, plaster walls damaged. Trees and bushes shook. Caving noticed on some steep hillsides and along stream banks.	
D. Felt by all, some had trouble walking. Not noticed by most drivers. Damage to structures slight, but windows broken, things knocked off shelves, pictures knocked off walls. Plaster walls cracked.	
E. Felt by most people. Water in swimming pools disturbed. Some reported broken glassware. Pendulum clocks stopped.	

3. The area of highest intensity was near Oakland and San Francisco (Fig. 16.12). How can the intensity be so high, so far from the epicenter? Might it be related to the type of sediment or rock at the surface? Figure 16.13 is a *Landsat* false color image of the San Francisco Bay area. Bedrock appears as reddish brown to dark brown tones. Mud and fill occur around the margins of San Francisco Bay, indicated by either nearly black, green, white, or bluish white tones. Between these two is a light blue area speckled with red, which represents unconsolidated natural alluvium.

   a. Outline on Figure 16.13 the areas in which these three types of materials occur around San Francisco Bay.

   b. Compare your results with Figure 16.14, a map showing the expected effects of ground shaking in the San Francisco Bay area and with the earthquake intensity map you drew. Which of the three types of material shook the most and why?

   c. Sketch hypothetical seismograms obtained from seismographs, located at Oakland, that sit on (1) mud and fill, (2) unconsolidated soil (mainly alluvium), and (3) stable bedrock. (Because of proximity to the epicenter, *P, S,* and surface waves will not be separated, so just show relative amplitudes of vibrations.)

   d. On what famous fault was the October 17, 1989, earthquake centered (see Fig. 16.14 for named faults)?

FIGURE 16.13
Landsat image of San Francisco Bay area for use with Problem 3.

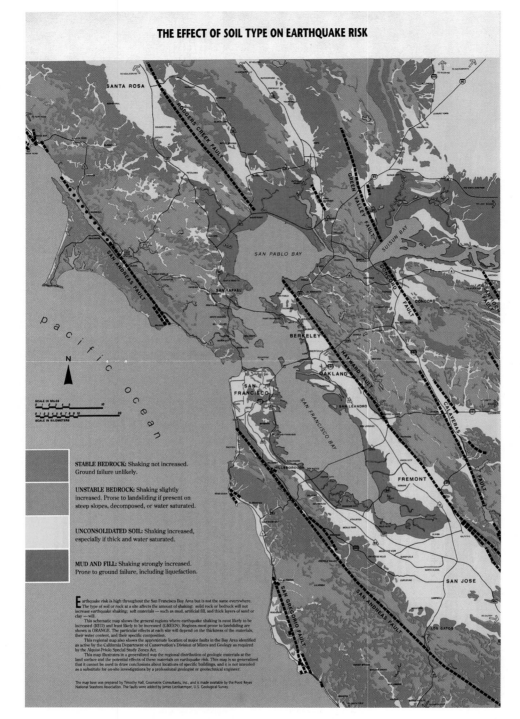

FIGURE 16.14
Map showing the expected effects of ground shaking in the San Francisco Bay area.

4. Go to *www.eas.slu.edu/Earthquake_Center/earthquakecenter.html* (or link to it through *www.mhhe.com/jones6e*—see Preface), select *Recent Earthquake Locations and Mechanisms*, then select *Recent Earthquake Maps*, and choose the map of the area in which you live (e.g., *Continental US and Adjacent Canada*).

   a. When was the map last generated?

   b. Where was the nearest earthquake in the previous fourteen days?

   c. What was its magnitude?

   d. What was the depth of focus?

   e. If you are not looking at the map *Continental US and Adjacent Canada*, return to *Recent Earthquake Maps*, and select that map. Which state had the largest number of earthquakes?

   Where was the largest earthquake on the map, and what was its magnitude?

5. **Past Earthquake Records.** Visit *www.ncedc.org/cnss/* (or link to it through *www.mhhe.com/jones6e*—see Preface) and select *Earthquake Maps & Lists*, then *Maps of Earthquakes*.

   a. In the first row (World Map Mollweide Projection) and eighth column (30 years of records), click on $M > 6.5$ to see the location of all earthquarkes with a magnitude greater than 6.5. Are shallow, intermediate, or deep focus earthquakes the most common? Which depth range is rarest?

   b. Around which ocean are most intermediate and deep earthquakes found?

   c. Now go to the Alaska Map row, 30 years of records column, and click on $M > 5.0$. Describe how the intermediate and shallow earthquakes are distributed relative to the string of Alaskan islands known as the Aleutian Islands and to each other.

   d. Go to the Western Hemisphere row, 30 years of records, and click on $M > 6.5$. The data are not as numerous here, and the view is from farther away, but can you see an earthquake focal-depth pattern analogous to that seen in Alaska? What is that pattern?

   Similar observations were important clues in the development of the theory of plate tectonics, as you will see in the next chapter.

You may want to check out *www.sciencecourseware.org/VirtualEarthquake/*. This is an excellent interactive site that leads you through the process of locating an earthquake and determining its Richter magnitude.

You also may find it interesting to look at the Global Earthquake Report by going to the *http://tsunami.geo.ed.ac.uk/quakexe/local-bin/quakes/mapscript/home.pl* site. There you find a list of the larger earthquakes that have occurred worldwide in the past week. You can access world and U.S. maps that show the epicenters, and you can zoom in on any area in those maps.

California
Aftershock at 02:07:00

Winnemucca, NV
P_1 arrival 02:08:36.07
P_2 arrival 02:08:36.94

P_1, 660 km

Crust

P_2 P_2

Mantle

FIGURE 16.15
Cross section illustrating two paths for P waves from aftershock focus in California to Winnemucca, Nevada, 660 km away. Times given in hour:minute:second, Universal Coordinate Time. For use with Problem 6.

In Greater Depth

6.   While the damage from earthquakes can be devastating, valuable information can be gained by studying them. The energy radiated from earthquakes can be used to look inside the Earth. An aftershock of the October 17 earthquake was recorded at Winnemucca, Nevada, a distance of 660 km away. Some of the P waves arriving in Winnemucca were reflected from the mantle, while others traveled there on a nearly direct path. Figure 16.15 illustrates, in cross section, two paths for the P waves. P_1 is the nearly direct path, and P_2 is reflected from the top of the mantle.

     a.   If the earthquake occurred at 02:07:00 (Universal Coordinate Time), use the arrival time of P_1 (and the distance of 660 km) to calculate the velocity of P_1; show your work.

     b.   Assume the velocity of P_2 is the same as P_1, and using its arrival time, calculate the distance traveled by P_2.

     c.   Half of this distance will be the hypotenuse of a right triangle such as triangle ABD in Figure 16.8. Use the Pythagorean theorem ($c^2 = a^2 + b^2$) to calculate the depth to the mantle. Show your work.

CHAPTER 17

Plate Tectonics

Materials Needed

- Pencil and eraser
- Metric ruler
- Protractor
- Calculator

Introduction

The plate tectonic theory is one of the most important and far-reaching theories in geology. It evolved from detailed observations made over many years and from individual hypotheses that required extensive and imaginative testing. For example, measurement of magnetism of the rocks on the seafloor was used as a test for seafloor spreading; the nature and distribution of earthquakes and volcanoes provided a test for plate convergence; and a variety of geological and geophysical studies proved the existence of transform faults. In this lab, a few predictions based on the plate tectonic hypothesis will be tested by analyzing appropriate data.

The theory of plate tectonics was more than 100 years in the making. Beginning in the 19th century, geologists began to compile basic observations about the Earth that could not be explained satisfactorily.

For example:

Fossils of the same kinds of land-dwelling plants and animals were found on opposite sides of oceans; how did these land dwellers cross the ocean?

Fossils of warm-weather plants and animals were found where climates are now cold, and evidence of glaciation was found in areas that now are warm; were Earth climates radically different in the past?

Earthquakes, volcanoes, and mountains are not randomly distributed about the Earth; why not?

Some interesting hypotheses were developed to explain these observations. To explain the distribution of fossils of land dwellers, it was supposed that continents were connected by land bridges that have since sunk into the oceans. To explain the rock-borne evidence of diverse climates, it was supposed that climates fluctuated widely in the geologic past. To explain the causes and distribution of earthquakes, volcanoes, and mountains, hypotheses involving contraction or expansion of the Earth were proposed; contraction would cause fold-and-thrust-faulted mountains, and expansion would cause fault-block mountains and provide conduits to the surface for magma. None of these hypotheses withstood testing, however, and eventually all were discarded.

The first detailed attempt at a unifying theory, one that could explain most of the major observations, was the continental-drift hypothesis of Alfred Wegener, proposed in a series of papers and books between 1910 and 1928. Wegener's hypothesis was severely criticized, especially by geologists in the northern hemisphere, on the grounds that it was mechanically impossible.

But more and more evidence seemed to suggest that continents indeed had moved relative to one another. Research that began in the late 1920s, and developed through the thirties and forties, indicated that a mechanism involving convection currents within the solid Earth might be possible. Additional geophysical evidence was especially persuasive, and by the early to mid-1960s, the framework of plate tectonics was in place.

At that time, plate tectonics was still a hypothesis, and there were many doubters, ready with observations that could not be readily explained by the existing model. The mark of a good hypothesis is that it can be tested, and the plate tectonics hypothesis offered many tests. Tests are conceived from predictions. That is, if the hypothesis is correct, then we can predict and test for certain consequent relations or results. In this lab, a series of predictions will be evaluated using real data.

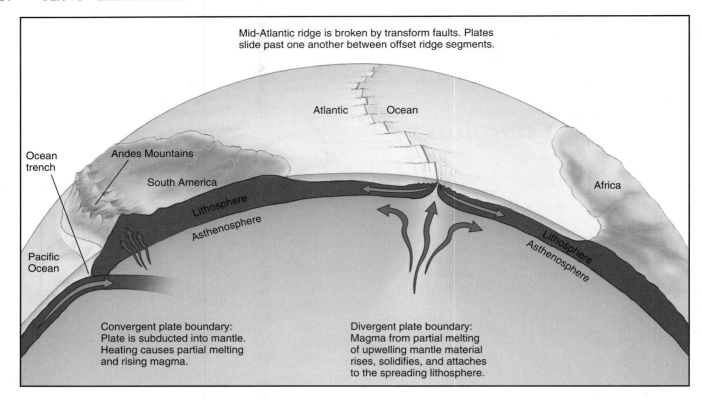

Mid-Atlantic ridge is broken by transform faults. Plates slide past one another between offset ridge segments.

Atlantic   Ocean

Ocean trench

Andes Mountains

South America

Lithosphere

Asthenosphere

Pacific Ocean

Africa

Lithosphere

Asthenosphere

Convergent plate boundary: Plate is subducted into mantle. Heating causes partial melting and rising magma.

Divergent plate boundary: Magma from partial melting of upwelling mantle material rises, solidifies, and attaches to the spreading lithosphere.

FIGURE 17.1

Cutaway view of the Earth, illustrating plates of lithosphere and the three types of plate boundaries. The thickness of lithosphere and asthenosphere is exaggerated.

Source: Modified from USGS, *Professional Paper 1350*.

General Theory of Plate Tectonics

According to the theory of plate tectonics, the outermost part of the Earth is made of **lithospheric plates,** some 70 to 125 km thick, that move relative to one another (Fig. 17.1 and inside front cover). Plates are constructed by successive injections of magma at **divergent plate boundaries.** The plates move away from the divergent boundaries and, at **convergent plate boundaries,** they encounter other plates moving toward them in a relative sense. There, one of the plates moves beneath the other, and either descends into the mantle or becomes sutured (attached) to the overlying plate. A **transform-fault plate boundary** is a type of strike-slip fault; plates slide past one another, and lithosphere is neither created nor destroyed. Let's look at these three types of plate boundaries in a little more detail.

Divergent Boundaries

Plates move apart at divergent plate boundaries as magma from the mantle wells up to fill the gap between adjacent plates, as shown in Figure 17.2. Divergent boundaries in the oceans are expressed by ridges and rises, generally with a narrow rift valley marking the boundary between plates along a central ridge crest.

Perhaps the most critical piece of evidence for the plate tectonic hypothesis, or at least for the seafloor-spreading component, came in the early 1960s from the study of the magnetic properties of rocks on the floor of the North Atlantic Ocean. This story has two parts.

The first has to do with the way in which an igneous rock retains a record (its **remnant magnetism**) of the existing magnetic field as it cools from a molten state. Re-

call that some minerals, especially magnetite (Fe_3O_4), are strongly magnetic. However, if they are heated above a certain temperature, the **Curie point** (about 580°C for magnetite), the magnetism disappears. Conversely, magnetite, which forms at about 1100°C as magma crystallizes, becomes magnetic when the temperature drops below its Curie point. Its magnetism aligns with the existing magnetic field, as shown in Figure 17.3A, and is frozen into the rock as though the rock contained tiny bar magnets. Thus, the rock contains a record of the magnetic lines of force at that time and place; the north-seeking ends of the "magnets" point toward magnetic north, and the axes of their magnetic fields are parallel to the **inclination** of the magnetic lines of force. The angle of inclination is determined by the magnetic latitude at which the rock formed (Fig. 17.3A).

The second part of the story has to do with reversals of the magnetic field and **magnetic anomalies** (magnetic values

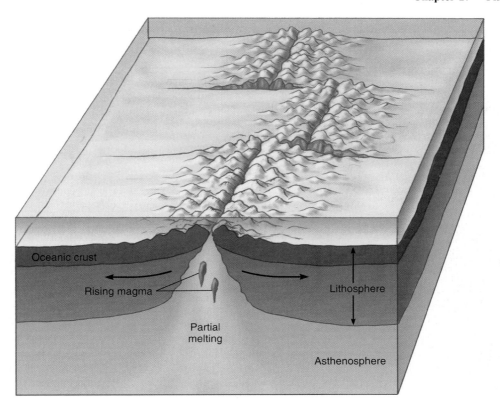

FIGURE 17.2
Block diagram of a mid-ocean ridge, a divergent plate boundary, at which seafloor spreading occurs.

greater or less than expected). In the present-day magnetic field, the north-seeking pole of a magnet points toward the north magnetic pole, which coincides approximately with the north geographic pole. But studies of remnant magnetism show that this has not always been the case. At many times in the geologic past, the orientation of the Earth's magnetic field has been reversed, as shown in Figure 17.3B. During these periods, the north-seeking poles of newly forming mineral "magnets" actually pointed toward the south geographic pole.

Imagine what could happen along a divergent boundary as magma rose, solidified, and took on the magnetic orientation existing at that time. During times of **normal magnetic polarity,** north-seeking remnant magnetic poles would point toward present-day north; during times of **reversed magnetic polarity,** north-seeking poles would point toward present-day south. As more magma is injected, the previously formed lithosphere (with one mag-

netic orientation) would be forced to move away from the plate boundary in both directions. Should the magnetic field reverse its polarity, newly injected magma would form rock that has a reversed polarity. Thus, as was hypothesized in the 1960s, alternating stripes or bands of rock with either normal or reversed remnant magnetism should form parallel to the divergent ocean ridges. Such magnetic stripes can be detected by a ship towing a magnetometer close to the seafloor. When passing close to rocks with normal remnant magnetism, their magnetic intensity is added to the modern magnetic field and a **positive magnetic anomaly** results. When passing over rocks with a reversed remnant magnetism, a weaker **negative magnetic anomaly** results.

It was an exciting time in geology when parallel bands of positive and negative magnetic anomalies were discovered along the Mid-Atlantic Ridge (Fig. 17.4). Here was a magnetic-tape-like record of past reversals of the magnetic field as recorded

by basalt extruded at a divergent plate boundary. If the numerical age of a sample from one of the bands on Figure 17.4 were known (for example, from radioactive isotopes in the sample), then the age of the entire band could be inferred as could that of the symmetrically equivalent band on the opposite side of the divergent boundary.

If seafloor spreading occurs as described, and if magnetic anomalies result from magnetic polarity reversals, you should be able to predict the answers to the following questions and propose ways to test your predictions. (1) Will magnetic anomalies on opposite sides of a divergent boundary correlate; that is, will a negative anomaly indicating a specific reversal occur on both sides of the boundary? (2) Will magnetic data from other ocean ridges or rises show the same general pattern of parallel bands as shown in the North Atlantic? (3) Will specific magnetic anomalies correlate on a worldwide basis; that is, will the same anomalies be found worldwide?

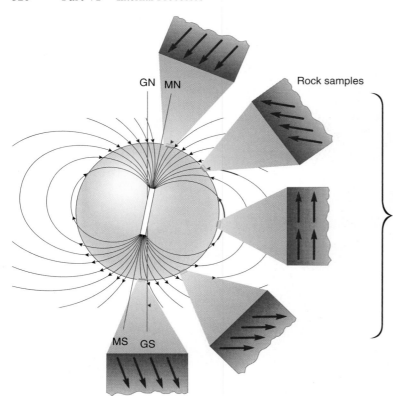

A. Normal geomagnetic polarity

Basalt samples from various latitudes have *normal* remnant magnetism. Arrows in samples are parallel to, and point in the same direction as, magnetic lines of force at each location and reflect the magnetic latitude at which rocks formed. Yellow marks upper surface of sample.

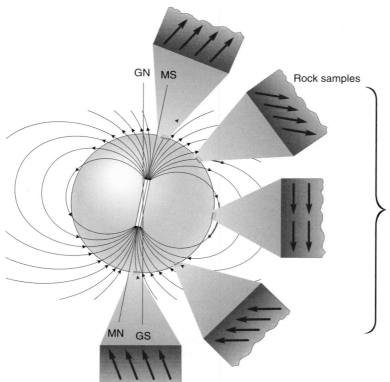

B. Reversed geomagnetic polarity

Basalt samples from various latitudes have *reversed* remnant magnetism. Arrows in samples are parallel to, and point in the same directions as, magnetic lines of force at each location and reflect the magnetic latitude at which rocks formed. Yellow marks upper surface of sample.

You also should be able to propose a way to use the magnetic anomalies to determine the rate at which seafloor spreading has occurred in a given area. Problems 1 and 6 deal with magnetic anomalies at divergent plate boundaries.

Convergent Boundaries

Plates come together at convergent boundaries, as illustrated in Figure 17.5. Converging plates contain either oceanic crust or continental crust at their leading edge. When two plates with oceanic crust converge, one dives, or is **subducted,** below the other (Fig. 17.5A). When an oceanic plate and a continental plate converge, the more dense oceanic plate is subducted beneath the continental plate (Fig. 17.5B). When two plates with continental crust collide, only partial subduction occurs. This is because continental crust contains rocks that are too buoyant (low density) to be forced down into the mantle. Oceanic rocks (basalt and gabbro) can subduct because they are only slightly less dense than the mantle. The partial subduction is accompanied by intense deformation and eventual suturing of segments of the subducted plate to the base of the overriding plate (Fig. 17.5C).

A subducted plate is comparatively cold and rigid when it begins its descent, but it warms and softens as it moves downward. Let's consider two of the many consequences implied by the converging process.

Earthquakes

Most of the world's major earthquakes take place on convergent plate boundaries. The

FIGURE 17.3
Earth's magnetic lines of force during (A) *normal* and (B) *reversed* geomagnetic polarity. During normal polarity, the magnetic north pole (MN) is near the geographic north pole (GN). When the magnetic pole is reversed, MN is near the geographic south (GS) and the magnetic south pole (MS) is near the geographic south. Lavas that cooled through the Curie point during normal polarity have a normal remnant magnetism with an orientation appropriate to the latitude at which they formed. This is illustrated in A by the red and yellow rock samples from various latitudes; samples with *reversed remnant magnetism* are shown in B.

FIGURE 17.4

Symmetrical magnetic anomalies about the Reykjanes Ridge portion of the Mid-Atlantic Ridge, southwest of Iceland. Dark areas are positive magnetic anomalies, light areas between them are negative anomalies. Spreading is perpendicular to the ridge axis.
Source: "Symmetrical magnetic anomalies about the Reykjanes Ridge portion of the Mid-Atlantic Ridge" from "Spreading of the Ocean Floor: New Evidence" by F. J. Vine in *Science,* Figure 2, Vol. 154, 1966, p. 107. Copyright © 1966 American Association for the Advancement of Science. Reprinted with permission.

zone along which most occur is named the **Benioff zone,** after seismologist Hugo Benioff. The faulting that generates the earthquakes is both along the boundary between the two lithospheric plates and within the descending slab itself (Fig. 17.5).

If convergence does take place as outlined, you should be able to predict the answers to the following questions, and propose ways to test your predictions. (1) Are earthquakes likely to be shallow (0–70 km), intermediate (70–300 km), or deep focus (300–700 km), or will there be a range of focal depths? (2) Will earthquake foci show the angle at which the plate descends? Problem 2 deals with these predictions.

Volcanic Arcs

Volcanic, or magmatic, arcs are associated with convergent plate boundaries. The volcanoes usually occur 100 to 300 km away from the trench that marks the position of the subduction zone on the surface, and are located on the same side of the trench as the earthquake epicenters. The implication is that magma is generated as a result of the subduction process.

Given these general observations and the hypothesis of plate tectonics, (1) suggest a process by which magma forms as a result of subduction; (2) predict how the **arc-trench gap** (the horizontal distance between the arc and the trench) may be related to the dip of the Benioff zone; and (3) predict whether subducted lithosphere must reach a certain depth for magma to form. Problem 2 relates to volcanic arcs.

Transform Faults

Transform faults are so called because they transform motion from one plate boundary to another. For example, mid-ocean divergent boundaries are broken into numerous segments by transform faults (Fig. 17.6). Transform faults are strike-slip faults; the principal movement on them is horizontal, and fault planes are essentially vertical. Like strike-slip faults, the sense of movement is described as right-lateral or left-lateral, depending on whether rocks on one side of the fault appear to have moved to the right or left, when viewed from the other side of the fault. Notice that the *apparent* sense of movement on a transform fault may be different than the *actual* sense of movement. For example, in Figure 17.6, the apparent offset of the ridges is right-lateral, but because of the divergence process, the actual movement is left-lateral, as shown by the arrows. Furthermore, movement is taking place only between the two segments of the offset ridge.

As with any other kind of fault, movement along a transform fault results in offset of features on opposite sides. Suggest:

1. how the total amount of movement on a transform can be determined from offsets, and

A.

B.

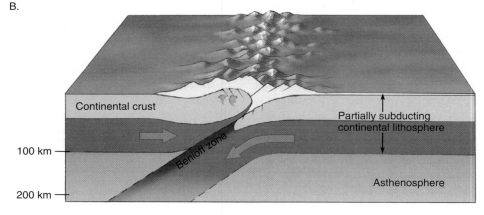

C.

FIGURE 17.5

Block diagrams of convergent plate boundaries. A. Both plates contain oceanic crust.
B. Subducting plate contains oceanic crust; overriding plate contains continental crust. C. Both plates contain continental crust. The subducted part of the slab in C is oceanic crust subducted before the continents collided.

2. how the amount of offset and the ages of the rocks offset can be used to determine the rate of movement and the time at which movement began.

    Problem 5 relates to transform faults.

Plate Motion and Hot Spots

At more than 100 places on Earth, plumes of hot material rise through the mantle and penetrate the lithosphere. These **hot spots** are sites of persistent volcanic activity. Some, like Iceland, coincide with present-day divergent boundaries, but many do not. Those hot spots within plates may produce strings of volcanoes that form as lithosphere drifts over them (Fig. 17.7). The volcanoes are active when directly over the hot spot, but die out as the lithosphere drifts on.

The position of these plumes within the mantle is fixed, which makes them useful in testing several aspects of plate-tectonic theory. For example, (1) predict how the ages of hot-spot-produced volcanoes should vary with distance from the hot spot, and suggest a way to test your prediction. (2) Suggest how hot-spot tracks (strings of volcanoes) could be used to monitor the direction of plate movement through time. Problem 3 relates to hot spots.

Beyond Plate Tectonics

In its basic form, plate-tectonic theory has rigid lithospheric plates drifting horizontally across the Earth's surface. Interactions between the plates cause intense deformation of their edges whereas the interiors remain largely undeformed. Mountains form when plates collide and pile up slivers of crust. Deep valleys and new oceans form when crust stretches, thins, and ultimately breaks. Plate tectonics is highly successful in explaining a huge number of geologic observations, but it does not explain everything. For example, why do some rare earthquakes occur in the middle of the great tectonic plates, far from any plate boundary? Another example: The Appalachians originally formed when Africa collided with North America some ~300 million years ago. However, erosion largely leveled the Appalachians by about 150 million years ago. Thus, the present-day mountains arose relatively recently, in the middle of a plate, independent of any collision. What caused this uplift? These and many other examples do not appear to invalidate the theory of plate tectonics, but they do show that there are still many things about the Earth that need better explanations.

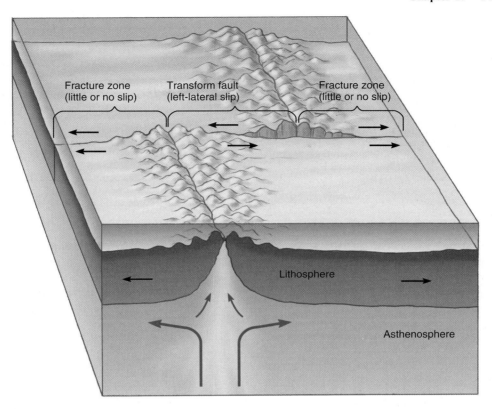

FIGURE 17.6
Transform fault joining segments of a mid-ocean-ridge, divergent-plate boundary. Movement on fault takes place between ridges but not on fracture-zone extensions.

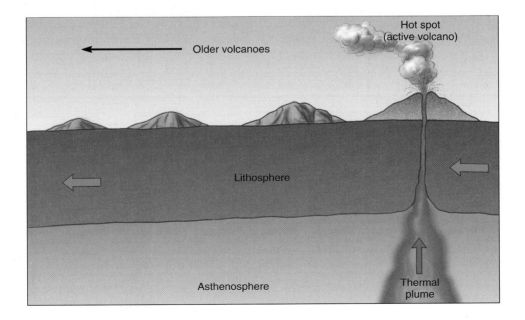

FIGURE 17.7
A fixed heat and magma source below the lithosphere causes a hot spot on the surface. As the lithosphere moves, a string of volcanoes is formed. Older volcanoes are lower and more rounded because they have undergone more erosion.

Hands-On Applications

The development of the plate-tectonic theory is an excellent example of how science works. As data accumulated and relations were established, various models or hypotheses were developed and tested. The following problems illustrate some of the methods used to test important aspects of the theory.

Objectives

If you are assigned all the problems, you should be able to:

1. Explain how a hypothesis is tested.

2. Explain how magnetic profiles across divergent-plate boundaries are interpreted to show divergence and magnetic polarity reversals.

3. Correlate magnetic profiles
   (a) along a divergent boundary and
   (b) with the geomagnetic polarity time scale.

4. Calculate divergence rates from magnetic anomalies, given their ages.

5. Draw a cross section of a convergent plate boundary that shows the Benioff zone, volcanic arc, trench,

arc-trench gap, forearc, and back-arc, and explain their distribution.

6. Use the ages and amount of offset of rocks or surface features to determine the average slip rate along a transform fault.

7. Explain how hot-spot volcanoes form, become extinct, and are used to determine speed and direction of plate movement.

Problems

1. Figure 17.8 shows magnetic profiles from data gathered in 1965 along four traverses in the South Pacific. The data were collected using a magnetometer (an instrument that measures magnetic intensity) towed behind a ship. Each profile crosses the Pacific-Antarctic Ridge, the crest of which is shown with a dashed line; water depth is 2 to 4 km. These data were among the first to test the hypothesis of seafloor spreading by comparing magnetic anomalies across and along ocean ridges. The peaks, or positive anomalies, on these profiles correspond to periods of normal magnetic polarity, and the valleys or negative anomalies correspond to times of reversed magnetic polarity.

   a. Four peaks that the authors of the original report found distinctive and useful are labeled on profile A. If the hypothesis of seafloor spreading is correct, similar peaks should be present at roughly equal distances from the ridge axis on the southeastern side of that profile. Identify peaks that are similar to 1, 2, 3, and 4 southeast of the ridge on profile A, and label them 1′, 2′, 3′, and 4′. In searching for matching peaks, use the surrounding peaks and valleys to help identify the numbered ones. To help you get started, 2′ is identified.

   b. A second requirement of the hypothesis of seafloor spreading is that the magnetic anomalies can be connected from profile to profile to form "stripes" parallel to the spreading axis (divergent-plate boundary). To test this, first find on profiles B and C the anomalies corresponding to the numbered ones on profile A. To help you, most are already located on profile D. Then connect those anomalies with the same numbers. You may have some difficulty locating the numbered anomalies. That's just the way it is—these are real data—so do your best.

   c. Explain how your results in *a* and *b* do or do not support the seafloor-spreading hypothesis.

   d. Peak 4 has been independently dated as being 7 million years old. What is the rate of seafloor spreading at this ridge? Calculate the *half-spreading rate* (at which one plate moves away from the ridge crest) and the *full-spreading rate* (at which points on the two plates move away from each other) using profile D. Give your answers in cm/yr.

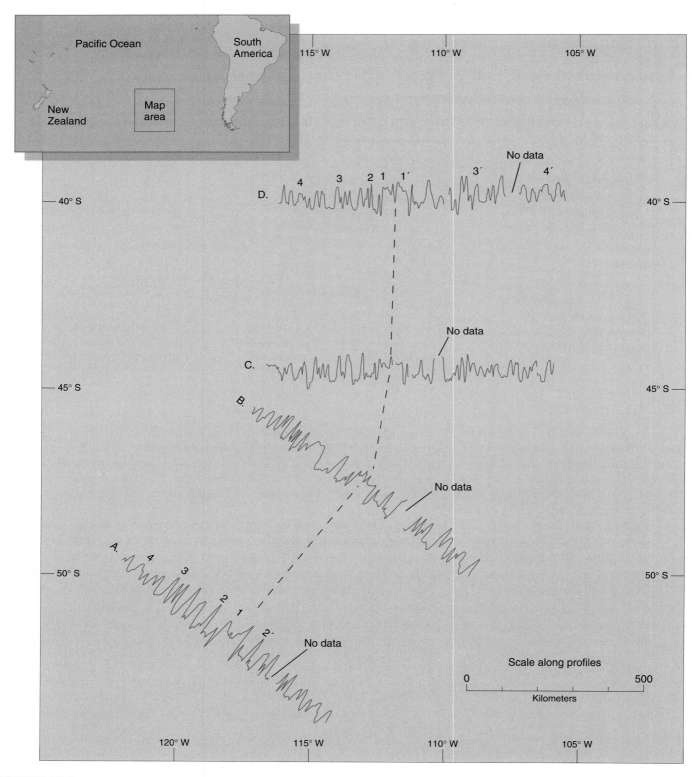

FIGURE 17.8

Four magnetic profiles across the Pacific-Antarctic Ridge. Dashed line marks the ridge crest. For use with Problem 1. Identify peaks 1–4 and 1′–4′ on each profile and connect them.

Source: "Four magnetic profiles across the Pacific-Antarctic Ridge" from W. C. Pitman, III, and J. R. Heirtzler in *Science,* 154:1164–71, 1966 Copyright © 1966 American Association for the Advancement of Science. Reprinted with permission.

EARTHQUAKE EPICENTERS

Depth (km)

28 Dec 1945 (7.8) ▼ ● ─┤70
10 Nov 1913 (7.5) ▼ ● ─┤200
02 Jul 1953 (7.5) ▼ ● ─┤300
23 May 1956 (7.5) ▼ ● ─┤400
● ─┤500
28 Sep 1957 (7.5) ▼ ● ─┤600
● ─┤700

─┐0

Focal depth Colored dots indicate 1964 through 1977 magnitude 5.0 through 7.4 events; triangles of matching colors labeled by date and magnitude indicate 1899 through 1985 events of magnitude 7.5 or greater (generally surface-wave magnitude)

VOLCANIC CENTERS

Active in historical time
Generally within past 1000 yr. and documented during or soon after eruption; volcano named in color if active 1964 through 1985

Active in Holocene time Within past 10,000 yr; not documented historically

Holocene activity uncertain
Includes volcano-related thermal springs and gas vents; no direct evidence for Holocene eruption

FIGURE 17.9

(Page 334) A plate boundary in the southwest Pacific Ocean near the Fiji, Samoa, and Tonga Islands. Shown on the map are bathymetry (water depth or depth to seafloor), active and recently active volcanoes, and epicenters of earthquakes of varying focal depths. Water depth is also shaded, with darker blue indicating deeper water. For use with Problem 2.

2. Figure 17.9 is a map showing a plate boundary in the southwest Pacific Ocean near the Fiji, Samoa, and Tonga Islands. Shown on the map are bathymetry (water depth or depth to seafloor, indicated by contour lines), active volcanoes, and epicenters of earthquakes of varying focal depths.

   a. What type of plate boundary is this? Given its location, what type of crust (oceanic or continental) do you think lies to the east? The west?

   b. The top part of Figure 17.10 is a topographic profile of the seafloor along line A-A′ in Figure 17.9. The topographic profile has a vertical scale of 1 cm to 3,125 m. What is the vertical exaggeration if the horizontal scale is 1 cm to 97.5 km?

   c. The graph on the bottom part of Figure 17.10 is for you to show, with no vertical exaggeration, the focal depths of earthquakes along the line A-A'. Plot generalized earthquake foci on that cross section by placing circles at the appropriate depths below areas in which earthquakes of various focal groups (different color codes) occur. Look north and south of the profile line to ensure your circles are *representative of the trend as a whole*. What is the significance of the focal-depth pattern on your graph; that is, why are the earthquakes where they are?

   d. What is the dip of the earthquake zone? (Measure with a protractor.) Draw arrows on the graph to illustrate relative motion along the zone.

   e. Compare the two profiles. What topographic feature is present where the zone of earthquake foci intersects the surface?

   f. Label the trench, forearc, volcanic arc, and backarc on the topographic profile. Figure 17.5 may be helpful. What is the width of the arc-trench gap? Why are there volcanoes here—how does the magma form? (Hint: see Chapter 3.) At what depth range does it happen?

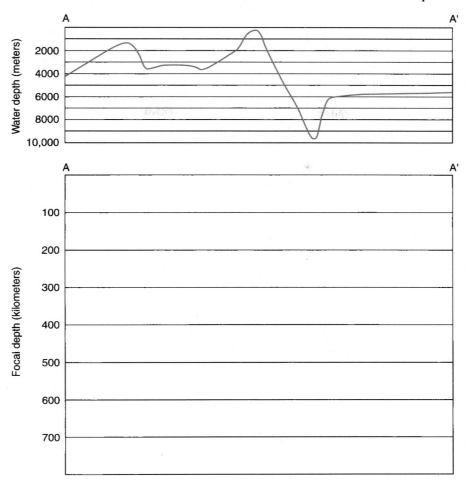

FIGURE 17.10
Section along line A-A′ of Figure 17.9, for use with Problem 2. A topographic profile appears on the top part of the diagram, and you can show earthquake foci on the bottom part. Horizontal scales are the same for top and bottom, but vertical scales are different.

3. Figure 17.11 is a map of the Hawaiian Islands and other islands and seamounts (submarine volcanoes) that form the Hawaiian-Emperor chain. All the features along the chain have a volcanic origin, and all are *younger* than the surrounding oceanic crust on which they sit. In 1963, J. Tuzo Wilson proposed that all the volcanoes in the Hawaiian chain had formed above the same hot spot or thermal plume. This idea was later extended to include the Emperor chain. If this hypothesis is correct, then (1) volcanoes should be older farther away from the hot spot, and (2) the distance-age relation can be used to measure the rate of plate movement.
    Table 17.1 lists the ages of these volcanoes and their distance from Kilauea, a currently active volcano on the Island of Hawaii.

    a. Plot the data in Table 17.1 on Figure 17.12, and label each of the points with the name of the volcano. What does the graph indicate about the general relation between age and distance from Kilauea (located at 0 km)? Draw two best-fit straight-line segments, the first between Kilauea and Laysan, the second between Laysan and Suiko. How do they differ?

    b. Calculate the rate of plate movement in centimeters per year for each of the straight-line segments: divide the distance traveled by the time interval over which travel took place. Remember to convert to the proper units. Are the two values the same? What does this mean? Show your calculations.

    c. The chain bends where the Hawaiian and Emperor chains meet. What could have caused the bend, and when did it happen?

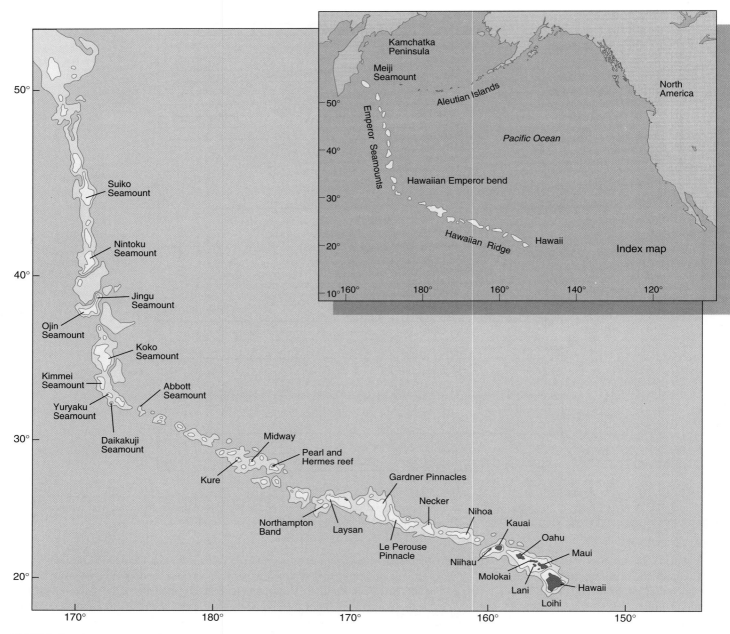

FIGURE 17.11

Hawaiian Islands and other islands and seamounts of the Hawaiian-Emperor chain, shown by 1- and 2-km bathymetric contours. For use with Problem 3.

Source: USGS, *Professional Paper 1350*.

Table 17.1
Selected Volcanoes of the Hawaiian-Emperor Chain
S o u r c e : USGS Professional Paper 1350.

Volcano Name	Distance from Kilauea (Km)	Age in Millions of Years		
Mauna Kea (on Hawaii)	54	0.375	±	0.05
West Maui	221	1.32	±	0.04
Kauai	519	5.1	±	0.20
Nihoa	780	7.2	±	0.3
Necker	1058	10.3	±	0.4
La Perouse Pinnacle	1209	12.0	±	0.4
Laysan	1818	19.9	±	0.3
Midway	2432	27.7	±	0.6
Abbott	3280	38.7	±	0.9
Daikakuji	3493	42.4	±	2.3
Koko (southern)	3758	48.1	±	0.8
Jingu	4175	55.4	±	0.9
Nintoku	4452	56.2	±	0.6
Suiko (southern)	4794	59.6	±	0.6
Suiko (central)	4860	64.7	±	1.1

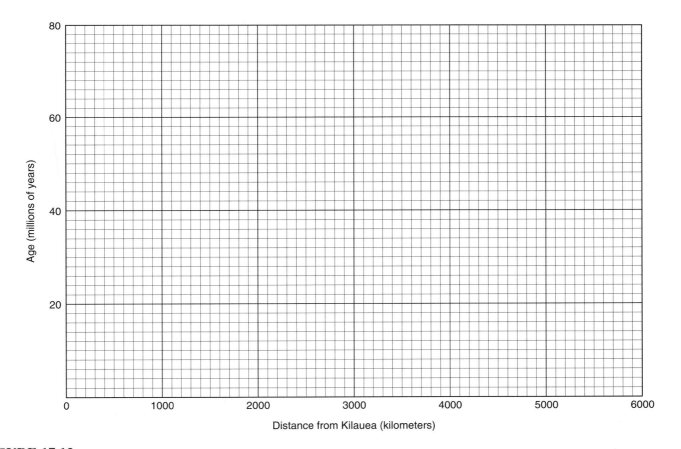

FIGURE 17.12
Graph of volcano age versus distance from Kilauea volcano (located at 0 km) along the Hawaiian-Emperor chain. Plot the data from Table 17.1, label points with the name of the volcano, draw best-fit straight lines between Kilauea and Laysan and between Laysan and Suiko, and answer the questions in Problem 3.

4. a. **Age of the Seafloor.** Go to *www.ngdc.noaa.gov/mgg/image/WorldCrustalAge.jpg* (or link to it through *www.mhhe.com/jones6e*—see Preface), which is a map showing the age of segments of the oceanic crust. Answer the following:

(1) Which areas have the oldest ocean crust (indicate along which coasts or what parts of a given ocean)?

(2) How old is the oldest crust? To what geologic period does this correspond?

(3) Where is the youngest oceanic crust, and what is its age? Be more specific, if you can, than the key at the bottom of the image.

(4) Explain the relation between the age of oceanic crust and its location in terms of plate-tectonic theory.

(5) Compare the age of the crust on the east and west sides of South America, and explain the age differences in terms of plate-tectonic theory. Hint: Look at a map that shows the tectonic plate boundaries (inside front cover).

(6) Compare the ages of the Atlantic oceanic crust along the coasts of Greenland and northern Europe (north Atlantic), the United States and Africa (central Atlantic), and Argentina and southern Africa (south Atlantic). Which part of the Atlantic Ocean opened up first, and which part last?

b. **Sediment on the Seafloor.** Go or link to *www.ngdc.noaa.gov/mgg/sedthick/sedthick.html,* which is a map showing the total thickness of sediment on the ocean floors. Answer the following:

(1) Where is sediment the thickest, near the continents or in the deep ocean? Why?

(2) The west side of South America is on a convergent plate boundary between a plate with oceanic crust and one with continental crust (e.g., Fig. 17.5B). There is an oceanic trench just offshore, and the Andes Mountains lie inland. The east side of South America is a trailing continental margin, which was once a divergent-plate boundary but now lies well within the South American plate.

(a) On which side of South America is sediment the thickest: the west side, adjacent to the high Andes Mountains, or the east side, where the relief is comparatively low?

(b) Formulate a hypothesis, based on plate-tectonic theory, to explain why this is so.

(3) Visit *http://mahi.ucsd.edu/Gabi/sediment.html* to see a more complete, but lower resolution, sediment-thickness map. Note the thickness of the sediment in the Bay of Bengal (east of India). This area is also close to a convergent plate boundary, but it is of a type like that in Figure 17.5C.

(a) How does the sediment thickness in the Bay of Bengal compare with that on the west side of South America?

(b) Formulate a hypothesis, based on plate-tectonic theory, to explain why this is so.

Also of interest is the U.S. Geological Survey site "Information on Plate Tectonics" *(http://geology.er.usgs.gov/eastern/ tectonic.html),* which contains several useful links.

In Greater Depth

5.   The San Andreas fault is a transform-fault boundary between the North America and Pacific plates in California. Figure 17.13 shows that it is really part of a network of faults collectively called the San Andreas fault *system*. The entire system accommodates movement between the two plates, although at present, most slippage occurs along the San Andreas fault itself. In this problem, you will learn how streams have been offset by the fault and how the study of offset streams and associated sediment can be used to determine the average rate at which movement has occurred in the recent past.

Figure 17.14 is an aerial view of the San Andreas fault in the arid Carrizo Plain of central California (see Fig. 17.13 for location). A stream, informally named Wallace Creek, drains the Temblor Range to the northeast (look up *temblor* in your dictionary) and crosses the San Andreas fault. The photograph and the map in Figure 17.15 show two prominent stream valleys, both of which bend sharply as they cross the fault. The map also shows three kinds of surficial deposits. It is the two stream-channel alluvium deposits that are of interest here, although fan alluvium and slope-wash deposits cover most of the area. The older-channel alluvium ranges in age from 10,000 yr BP (years before present, with "present" defined as 1950) to 3680 yr BP. The ages are based on radiocarbon (^{14}C) dates from pieces of charcoal and organic matter found in the deposits.

a.   Why does the present-day stream channel bend sharply at the San Andreas fault? Can you think of two *possible* explanations? Which is much more likely, and why?

b.   Based on your answer to *a,* explain the map pattern of the older-channel alluvium.

c.   What evidence is there on the map or photograph for *vertical* movement on the San Andreas fault?

d.   The following statement was made in an article describing Wallace Creek: "Mr. Ray Cavanaugh, who farms at Wallace Creek, reported to us that water actually spilled over the edge [where Wallace Creek bends sharply northwest] in the winter of 1971–1972 or 1972–1973." Based on this statement and your answers to *a* and *b,* predict the future of Wallace Creek. (Note the orientation of the north arrow in the *Explanation* in Figure 17.15.)

e.   Movement along the San Andreas fault in this area is by "fits and starts." The last large earthquake here, in 1857, was the result of 9.5 m of instantaneous slip. The *average slip rate* of the fault can be determined for two periods in the Wallace Creek area, because two channels of Wallace Creek are displaced by the fault, and the time when the channels first formed is known.

Calculate the two slip rates by measuring the offset across the fault for both channels and dividing by the greatest age of the channel deposits. Give your answers in centimeters/years. Show your calculations. Note: The map shows that the channel deposits in both channels just southwest of the fault line "bulge" toward the northwest, where the channels curve abruptly west. The deposits do this because the channels were eroded at the bends before all of the channel alluvium was deposited. Therefore, to get a better idea of the position of the channel when the oldest alluvium was deposited, use points farther downstream for your measurements.

f.   The best estimate of the total displacement along the San Andreas fault is about 320 km. This estimate was determined by matching distinctive groups of rocks that are offset by the fault. If the average slip rate you just calculated has been the same for the life of the fault, how old is the San Andreas fault? (Show your work, and remember to keep units consistent.)

FIGURE 17.13
The San Andreas fault system. Red dot is location of Wallace Creek.
Source: USGS, *Professional Paper 1515*.

FIGURE 17.14
Oblique aerial photograph of area shown on map in Figure 17.15.
The modern and abandoned channels of Wallace Creek are visible
on the foreground side of the San Andreas fault.

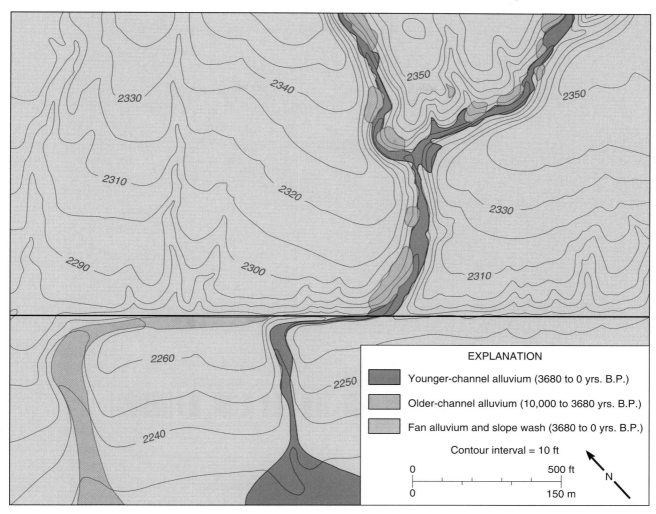

FIGURE 17.15
Map of Wallace Creek area, for use with Problem 5.
Source: Map of Wallace Creek from *GSA Bulletin,* Vol. 95, No. 8, 1984. Copyright © 1984 by Geological Society of America. Reproduced with permission of Geological Survey of America via Copyright Clearance Center.

6. Figure 17.16 is an enlargement of profile A of Figure 17.8. If the hypothesis is correct—that peaks on the profile represent normal magnetic polarity, and valleys, reverse magnetic polarity—then peaks and valleys should match periods of normal and reversed magnetism known from well-dated rocks from other parts of the world.

    a. To test this, first fill in the bar above the profile by blackening the areas corresponding to peaks. The dark areas corresponding to large peaks should have widths equal to the width of the peak, as measured about halfway between the top and the bottom. Dark areas for the small peaks should be lines. Several of the trickier ones have been filled in to help you get started, and a few ambiguous parts of the profile have been simplified.

    b. Compare your bar diagram with the geomagnetic polarity time scale in Figure 17.17. Start at time zero and see if the patterns match, going back in time. If you are able to match the two, mark points on your diagram (Fig. 17.16) that correspond to time intervals of one million years.

    c. Compare the width (i.e., distance in km) of the spaces between the million-year points you marked above. Has the half-spreading rate varied over time? Are half-spreading rates for the same time interval equal on both sides of the spreading center?

FIGURE 17.16
Enlargement of Profile A from Figure 17.8, for use with Problem 6.

| Magnetic polarity (black = normal, white = reversed) | |
| Age (millions of years) | 1   2   3   4   5   6   7   8   9   10 |

FIGURE 17.17
Geomagnetic polarity time scale.

Credits

Part Openers
One: © McGraw-Hill Companies/Doug Sherman, photographer; Two: © Doug Sherman/Geofile; Three: U.S. Geological Survey, Reston, VA; Four: Courtesy of Earth Satellite Corporation (EarthSat); Five: © Norris W. Jones; Six: © Doug Sherman/Geofile.

Chapter 1
Figures 1.1, 1.2A,B: © Doug Sherman/Geofile; 1.2C,D © Charles E. Jones; 1.2E: © Doug Sherman/Geofile; 1.2F: © The McGraw-Hill Companies/Dr. Parvinder Sethi, photographer; 1.2G,H, 1.3, 1.5, 1.6B, 1.7A, 1 9: © Doug Sherman/Geofile; 1.10(left): © Charles E. Jones; 1.10(right), 1.11(both): © Doug Sherman/Geofile; 1.12: © Charles E. Jones; 1.13: © Doug Sherman/Geofile; 1.14A,B: © Charles E. Jones; 1.15, 1.16, 1.17A,B, 1.18A,B, 1.19, 1.20, 1.21: © Doug Sherman/Geofile.

Chapter 2
Figures 2.1A-C, 2.2: © Doug Sherman/Geofile; 2.3: © Charles E. Jones; 2.4A-E: © Doug Sherman/Geofile; 2.4F, 2.5: © Charles E. Jones; 2.6, 2.7, 2.8: © Doug Sherman/Geofile; 2.9A: © Charles E. Jones; 2.9B, 2.10, 2.11, 2.12: © Doug Sherman/Geofile; 2.13: © Charles E. Jones; 2.14, 2.15, 2.16, 2.17: © Doug Sherman/Geofile.

Chapter 3
Figures 3.1A,B, 3.2A,B, 3.3, 3.4, 3.5, 3.7A,B, 3.8, 3.9A,B, 3.10A,B: © Doug Sherman/Geofile; 3.11A: © Charles E. Jones; 3.11B: © Doug Sherman/Geofile; 3.12A: © The McGraw-Hill Companies/Dr. Parvinder Sethi, photographer; 3.12B: © Doug Sherman/Geofile; 3.13A: © The McGraw-Hill Companies/ Dr. Parvinder Sethi, photographer; 3.13B: © Doug Sherman/Geofile; 3.14: © Charles E. Jones; 3.15A,B, 3.16A,B, 3.17: © Doug Sherman/Geofile.

Chapter 4
Figure 4.3: © Doug Sherman/Geofile; 4.4A,B, 4.5B: © Norris W. Jones; 4.6, 4.7: © The McGraw-Hill Companies/Dr. Parvinder Sethi, photographer; 4.9, 4.11, 4.12A,B, : © Doug Sherman/Geofile; 4.13A © The McGraw-Hill Companies/Dr. Parvinder Sethi, photographer; 4.13B, 4.14A,B © Doug Sherman/Geofile; 4.15A-1: © The McGraw-Hill Companies/Dr. Parvinder Sethi, photographer; 4.15A-2, 4.15B, 4.16A-E, 4.17, 4.18, 4.19, 4.20, 4.21: © Doug Sherman/Geofile.

Chapter 5
Figure 5.3A: © Doug Sherman/Geofile; 5.3B: © The McGraw-Hill Companies/ Dr. Parvinder Sethi, photographer; 5.4, 5.5, 5.6, 5.7, 5.8, 5.9, 5.10A,B, 5.11A: © Doug Sherman/Geofile; 5.11B: © Charles E. Jones.

Chapter 6
Figures 6.1A, 6.6, 6.9, 6.10, 6.15, 6.17: U.S. Geological Survey; 6.20: Courtesy of Natural Resources Canada.

Chapter 7
Figure 7.2A,B: NASA/GSFC/MITI/ERSDAC/ JAROS, and U.S. Japan ASTER Science Team; 7.4A,B: U.S. Geological Survey, ESIC; 7.5: U.S. Geological Survey/photo by D.R. Crandell; 7.6: © Dr. Juerg Alean; 7.7: U.S. Geological Survey/Photo by R.P. Hoblitt; 7.8: National Park Service/Photo by M. Williams; 7.9: U.S. Geological Survey/Photo by R.W. Decker; 7.10: U.S. Geological Survey/Photo by Peter W. Lipman; 7.11: U.S. Geological Survey/Photo by Steve R. Brantley; 7.12: U.S. Geological Survey/Photo by R. Clucas; 7.13: Created by Charles E. Jones from publicly available USGS data; 7.14: Hoblitt and others (1998) USGS Open-File Report 98-428; 7.15: U.S. Geological Survey, ESIC; 7.16, 7.17: U.S. Geological Survey, Denver, CO; 7.18: NOAA; 7.19: NASA; 7.20 NASA/JPL; 7.21: NASA/GSFC/MITI/ERSDAC/JAROS, and U.S. Japan ASTER Science Team; 7.22: NOAA.

Chapter 8
Figures 8.3, 8.4, 8.10: U.S. Geological Survey, Rolla, MO; 8.11: U.S. Geological Survey, Denver, CO; 8.12, 8.13, 8.14: U.S. Geological Survey, Rolla, MO; 8.16: U.S. Geological Survey, Denver, CO.

Chapter 9
Figures 9.8, 9.9: U.S. Geological Survey, Rolla, Mo; 9.10: Courtesy of James F. Quinlan and Joseph A. Ray, 1981; revised 1989.

Chapter 10
Figure 10.1: Austin Post, U.S. Geological Survey, Tacoma, WA; 10.2: Robert Felder, U.S. Geological Survey, Menlo Park, CA; 10.4: © Norris W. Jones; 10.7: U.S. Geological Survey, Denver, Co; 10.8: Prepared from USC & GS Photography by the University of Illinois Committee on Aerial Photography; 10.9: U.S. Geological Survey; 10.10: The Geological Society of America; 10.11: U.S. Geological Survey, ESIC; 10.12: U.S. Geological Survey, Rolla, MO; 10.14: USGS prof. Paper 1180, 1980, Courtesy U.S. Geological Survey, Denver, Co.

Chapter 11
Figure 11.8: © Doug Sherman/Geofile; 11.9A-C: Prepared from USC & GS Photography by the University of Illinois Committee on Aerial Photography; 11.10, 11.11, 11.12: U.S. Geological Survey, Denver, CO; 11.13A,B: U.S. Geological Survey, NAPP.

Chapter 12
Figures 12.1, 12.2, 12.5, 12.6: © Doug Sherman/ Geofile; 12.9: Prepared from USC & GS Photography by the University of Illinois Committee on Aerial Photography; 12.10, 12.11, 12.12, 12.13: U.S. Geological Survey, Denver, Co.

Chapter 13
Figure 13.3: © Norris W. Jones.

Chapter 14
Figure 14.2: © Norris W. Jones; 14.3A: © The McGraw-Hill Companies/ Dr. Parvinder Sethi, photographer.

Chapter 15
Figure 15.2B: U.S. Geological Survey, ESIC; 15.3B: Prepared from USC & GS Photography by the University of Illinois Committee on Aerial Photography; 15.4B, 15.5B, 15.6B: U.S. Geological Survey, ESIC; 15.12B,C: U.S. Geological Survey; 15.14: N.H. Darton 1951. Geologic Map of South Dakota, U.S. Geological Survey; 15.15: U.S. Geological Survey, EROS Data Center; 15.17B, 15.18B U.S. Geological Survey.

Chapter 16
Figure 16.6: U.S. Geological Survey, Denver, CO; 16.13: U.S. Geological Survey, EROS Data Center; 16.14: Timothy Hall, U.S. Geological Survey.

Chapter 17
Figure 17.9: American Association of Petroleum Geologists; 17.14: U.S. Geological Survey, Menlo Park, CA.

Index

Note: An *f* or *t* following a page number indicates a figure or table that appears outside the related text discussion.